SMOKING EARS
and
SCREAMING TEETH

Trevor Norton is an Emeritus Professor at the University of Liverpool, having retired from the Chair of Marine Biology. He has published widely on ecological topics. He is also an Honorary Senior Fellow at the Centre for Manx Studies on the Isle of Man where he lives. His much acclaimed books include *Stars Beneath the Sea*, *Reflections on a Summer Sea* and *Under Water to Get out of the Rain*.

Praise for Trevor Norton

'We are gently but passionately introduced to a magical world . . . Here is the joyous record of a fulfilled life.' *Times Literary Supplement*

'Brilliant . . . Norton's passion comes to life with his superb writing style. He's evocative, funny and thought provoking – this is the summer's best read.' *Choice*

'A sumptuous memoir. Norton's skill with language . . . infuses this book with a sweet and tranquil beauty.' *Boston Globe*

'Entrancing . . . A wonder-filled experience.' *BookBrowse* guide to exceptional books

'A witty and engaging memoir . . . precisely echoes the style of naturalist Gerald Durrell.' *Booklist*, American Library Association

'Norton's delightful sea shanty.' *The Independent*

'Eloquent, easy to read, witty and informative. I have seldom read a natural history book that was as delightful and enjoyable as this and never one that actually made me laugh out loud.' American Association for the Advancement of Science

Praise for *Reflections on a Summer Sea* by Trevor Norton

'A lovely book – an affectionate portrayal of rural Ireland in the sixties and a richly nostalgic memoir . . . Norton writes beautifully.' *Sunday Express*

'Fascinating . . . You only have to open *Reflections on a Summer Sea* to experience how thrilling the study of natural history can be . . . I thank him for making me feel young again.' David Bellamy

'Wonderful. Norton has a rare ability to recapture those ecstatically funny moments, and a profound and quite moving ability to capture what it is to be human. It's a world . . . immortalized in his crystalline prose. This is a book that you don't ever want to end and having read it you want to go straight back and start all over again.' Frank Ryan, winner of the *New York Times* Non-fiction Book of the Year

'A fine and touching tale. Highly recommended.' *The Bookseller*

'A wonderfully warm, gentle book. The richness of Irish life is intermingled with the wonders of natural history.' *Book of the Month*

'A delightful, lyrical, funny book, tinged with sadness that you just want to go on reading.' *Choice*

'Laden with wit and wisdom. Wonderful images remain in the mind. Norton's writing is magical.' *Irish Southern Star*

Praise for *Stars Beneath the Sea* by Trevor Norton

'Norton has shown that a gifted writer is an alchemist . . . His agile prose is burnished with humour and he has the natural storyteller's eye for detail.' *Daily Telegraph*

'A quirky history of the eccentric experiments of some truly mad individuals. I loved it.' Choice of the Month, *The Bookseller*

'Norton delights infectiously . . . and writes with wit and a fine eye for the poetry in the scientific work . . . His narrative is by turns funny and gripping.' *Guardian*

'A marvellous book . . . from the first page you are snatched into a racing current of excitement, adventure and discovery . . . He writes with humour too.' *New Scientist*

'Delightful. Norton writes with a light touch and a wonderful feel for his material.' *Publishers' Weekly*, New York

'Absolutely delightful.' *The Good Book Guide*

'Rich entertainment.' *Mail on Sunday*

'Trevor Norton's entertaining history of diving.' Bill Bryson

'Admirably concise and witty, entertaining and informative.' *Booklist*, New York

'Gripping, informative, often amusing, often sad, always interesting.' *Newslink*

'Truly magical.' David Puttnam

SMOKING EARS

and

SCREAMING TEETH

TREVOR
NORTON

arrow books

Published in the United Kingdom by Arrow Books in 2011

1 3 5 7 9 10 8 6 4 2

First published in the United Kingdom in 2010 by Century

Arrow Books
The Random House Group Limited
20 Vauxhall Bridge Road, London, SW1V 2SA

Addresses for companies within The Random House Group Limited can be found at:
www.randomhouse.co.uk/offices.htm

The Random House Group Limited Reg. No. 954009

www.rbooks.co.uk

A CIP catalogue record for this book
is available from the British Library

ISBN 9780099533597

Penguin Random House is committed to a sustainable future for
our business, our readers and our planet. This book is made from
Forest Stewardship Council® certified paper.

Printed and bound in Great Britain by Clays Ltd, Elcograf S.p.A.

Typeset by Palimpsest Book Production Limited,
Falkirk, Stirlingshire

To Charlotte and Katie, two of the most delightful distractions while trying to write a book. And Win to whom I read my latest chapters in bed while she lovingly fell asleep.

Acknowledgements

I am grateful to the Directors of the Centre for Manx Studies for taking in 'the man who orders weird books' after my retirement. I have benefited greatly from access to the libraries of Liverpool University and the Liverpool School of Tropical Medicine and I owe a special debt to Christine Sugden and her staff at the medical library, Keyll Darree, DHSS Education & Training Centre in the Isle of Man.

Thanks are due to those who drew my attention to literature I might otherwise have missed: Lynn Delgaty, Bernard Eaton, Andrew Sigley, Reg Vallintine, and Jon Franklin who kindly sent me the typescript of his book.

I am indebted to those who facilitated my access to information and people, particularly Dr John Bevan, Dr Andrew Brand, Drs Terry and Selma Holt, Mark Potok, Jenny and Gill at CMS, Erik Ahlbom for his translation skills, and the late and much missed Rosemary Pickard, and her staff at the Bridge bookshop.

Sincere thanks to Rachel Norton Buchleitner and Nick Austin for their meticulous proofreading, James Hamilton-Paterson for his helpful views on the book before I began to write it, Anna Webber for her encouragement, and Mark Booth and Charlotte Haycock who at all times acted as if my strange idea for a story was perfectly normal.

As ever, I have my wife Win to thank for many of the vignettes that adorn the chapters.

Preface

'*Live dangerously*' – Friedrich Nietzsche

Scientists are curious in all senses of the word. I have spent my life experimenting, but only once on myself. I had read that *urinatores,* the frogmen of ancient Rome, went diving with a mouthful of oil. No one knew why. Perhaps they cupped their hands over their eyes and dribbled out the oil to form the oil-bubble equivalent of a lens so that they could see underwater. I determined to put it to the test so I submerged with a large bottle of cooking oil. After repeated attempts I saw nothing, swallowed half a bottle of Mazola and had the runs for a week.

When schoolchildren were asked to draw a scientist, ninety per cent of them drew a *mad* scientist. The incidence of self-experimentation by researchers may seem to justify this description. In the name of science they have swallowed tumblers of cholera, hydrochloric acid and some unmentionable things that I will be sure to mention.

Why would they do such things? It's a strange tale of altruism, vanity, courage, curiosity and − of course − stupidity.

'If a little knowledge is dangerous, where is the man who has so much knowledge as to be out of danger?' − Thomas Huxley

He came, he sawed, he chancred

'Speak with caution of what may be passing here, especially with respect to dead bodies' — William Hunter

In the eighteenth century medical men were either cultured physicians well-versed in the theory of medicine, or surgeons, practical men with saws. Both were steeped in ancient lore and received wisdom. Medical research was stagnant and patients were little better off than their great-great-grandparents had been. Then along came a Scottish farmer's lad called John Hunter who changed surgery from a trade into a science.

John's education was basic, but he had an unquenchable curiosity for nature that remained with him throughout his life. In 1748 he departed for London to join his elder brother William. Although he had trained as a surgeon, William sometimes fainted at the sight of blood so he was changing careers to become a fashionable physician and male midwife. John was enlisted to handle the bloody side of the business and was set to work preparing cadavers for teaching purposes. His skill at dissection was astonishing and he soon graduated to supervising William's students. After being apprenticed briefly to two famous surgeons he became a house surgeon at St George's Hospital, which had been set up to treat the 'deserving poor'. It also gave surgeons licence to practise on the *uncomplaining* poor. Those who unknowingly offered their bodies for the training of surgeons were the vulnerable and the beneficiaries were mostly the wealthy. John spent his mornings visiting paying clients, and devoted the afternoons to treating the poor for no fee. At St George's he attracted more poor patients than all the other surgeons combined.

John hoped that the hospital would give more emphasis to educating young surgeons, but failed to persuade the senior surgeons to give lectures. Eventually he gave evening lectures in his own home and over the years these became the inspiration for all young medics in London. They were well attended although on one occasion only a single student turned up. To augment the audience John hauled in a skeleton and began with his usual opener – 'Gentlemen.'

John Hunter never blindly followed current practice: he always observed, then improved. A stint as an army surgeon during the Seven Years War made him an authority on gunshot wounds. Battlefield surgery involved opening the wound to scrape out any debris and extract the bullet. Almost invariably the soldier died from an infection. Hunter achieved a much higher survival rate by simply staunching the blood and leaving the bullet in place. He learned that the human body could sometimes heal itself.

He dissected over a thousand corpses and knew the interior of the human body better than the layout of his own house. The more he knew, the fewer surprises there would be on the operating table. The acuteness of his mind matched the dexterity of his hands. He came to know not just the parts of the body, but 'their uses in the machine, and in what manner they act to produce the effect'.

Hunter was not alone in his obsession; the artist George Stubbs spent eighteen months dissecting horses, working on each carcass for weeks. The rank smell would have turned the stomach of a less determined man, but the end result was his meticulous and monumental treatise on the anatomy of the horse.

John's brother William founded a private medical school in Great Windmill Street, where almost two centuries later the public's interest in anatomy would be satisfied by London's first strip club. William's aim was to provide the practical anatomical skills neglected elsewhere. Medical examinations, even for surgeons, were usually verbal affairs with no practical test whatsoever. Most courses taught surgery with students witnessing a dissection, or examining sample dissections prepared earlier. A parsimonious Scottish professor made a single cadaver last for an entire course of a hundred lectures. In those days it wasn't just the students that got high.

The first cut that an aspiring surgeon made might well be on a live patient. Both William and John believed that surgical mistakes were best made on the dead, not the living. John taught his students that: 'Anatomy is the basis of surgery, it informs the head, guides the hand and familiarises the heart to a kind of necessary inhumanity.'

In Hunter's school each student would have a corpse of his own to practise on. That meant a lot of bodies and they had to be fresh, although not fresh as Tesco and Sainsbury's know it. Dissection was largely a winter activity. Summer warmth rapidly dried the skin of a corpse as stiff as wood and turned its internal organs into glutinous porridge.

The school needed several bodies per week and John was given the task of finding them. There were over two hundred crimes – including pickpocketing – that carried the death penalty. The Murder Act of 1752 allowed anatomists to claim the bodies of executed murderers, so the surgeons converged on the gallows at Tyburn Tree. The bodies had to be warm from the scaffold and this led to unseemly tugs-of-war with the relatives of the deceased. One tussle was so vigorous that it revived the felon, who was reprieved and rechristened 'Half-hangit Maggie'.

The new provision of the Murder Act was not designed to help the progress of medicine but to punish criminals.

Being dissected was feared as a fate worse than death; it added 'a further terror and peculiar mark of infamy'. Also, on Judgement Day when all the dead would rise again, some would reappear with important pieces of themselves missing and might be refused admission into Paradise on the grounds of their incompleteness.

For many criminals the real fear was that they might awaken on the dissecting table with their entrails out on display. In the days before 'long-drop' hanging that broke the neck the condemned were slowly strangled by the noose – sometimes it took thirty minutes or more. In the melee that followed, the doctor often had no opportunity to pronounce the victims dead and there were several examples of their 'corpses' sitting up under the surprise of the anatomist's knife.

With an understandable dearth of volunteers there was no legal way to acquire sufficient cadavers, so anatomists often had no choice but to bribe undertakers to put stones in the coffin and hand over the dead loved one. Gravediggers were also obliging as it was no more trouble to bury an empty casket than a full one. Even these means could not meet demand so John Hunter began 'hobnobbing with the resurrection men' – body-snatchers who dug up the recently deceased.

During Hunter's working life grave robbing grew from the occasional 'uplifting' experience to resurrection on a scale to rival the Day of Judgement. Cadavers were being supplied to order and transported all around the country in hampers and barrels. A body in a box was found on the stagecoach heading for Leeds. A similar incident in Dublin caused the local paper to request that 'for the sake of decency, they packed their treasures a little more carefully'. The price rose sixteen-fold and children's bodies were charged for by the inch. Some 'Sack 'em up' gangs complained of the frequency with which they dug up a coffin only to find it had already been vacated.

Grave robbing was not a criminal offence. Stealing a pig or a goose was punishable by death, but in the eyes of the law a body was not property and therefore couldn't be stolen. The body-snatchers were careful to leave the shroud and clothes behind in the coffin because they *were* property.

The public were alarmed and there were violent anti-anatomist riots from Carlisle to New York. A medical journal protested that 'If the traffic in human flesh be not prevented, the churchyards will not be secure against the shovel of the midnight plunderer, nor the public against the dagger of the midnight assassin.'

In the interests of ensuring freshness, some criminals streamlined the process by snatching the body before it was dead. The papers were full of the escapades of Burke and Hare who murdered sixteen people and sold them to Edinburgh surgeon Robert Knox.

> Burke's the butcher, Hare's the thief,
> Knox the man who buys the beef.

When the murdered body of 'Daft Jamie', a well-known local character, was brought in, Knox decapitated the corpse before giving it to the students lest the victim be recognised.

Burke and Hare spawned a tribute gang, 'The Burkers', who supplied the still-warm to King's College London. The ringleaders were convicted of multiple murders and for 'resurrecting' a thousand newly buried bodies. The scale of this scandal led to the Anatomy Act of 1832 that gave anatomists the right to dissect all unclaimed bodies from workhouses and morgues. Now the fate that terrorised the most hardened rogue was visited on the innocent poor.

To Hunter, acquiring dead bodies was a surgical necessity that in the long run would save lives. It was also a game. The President of the Royal College of Surgeons boasted to a Royal Commission that: 'There is no person, let his situation in life be what it may, whom, if I were disposed to dissect, I could not obtain.'

The problem of too few cadavers is not entirely buried in the past. Today anatomy teaching involves anatomical dummies, medical imaging techniques and students being encouraged to examine their own bodies and those of others.

I seem to recall that students always examined each others' bodies at every opportunity. Even so, the classes at Guy's, King's and St Thomas' Hospitals need eighty bodies a year and only about sixty are donated. The dilemma is not helped by rejecting those that are obese, which will soon exclude everybody.

Most hospitals do their best to encourage people to donate their bodies, but some contrive to do the opposite. In 2004 a Californian medical school was found to have sold parts from around eight hundred corpses donated for medical research over a six-year period. A university in Louisiana had disposed of surplus cadavers to a broker and had no idea where they ended up.

The trade in organs is worth over one billion US dollars a year. Heart, lung or liver transplants are only a minor portion of it. The increasing longevity of people in the developed world has fuelled the demand for replacement parts such as corneas for failing eyes and bones for worn joints, not to mention skin grafts for burns, tendons and ligaments to mend injured athletes and collagen to make lips pout.

In Britain there were almost two thousand kidney transplants in 2007, but that left 8,600 patients on the waiting list for new kidneys. Over a thousand people each year die waiting for an appropriate organ to become available. The shortage of donors pushes up the price of parts. If fresh and sold separately, the marketable parts from a single cadaver can fetch $200,000; a head sells for $900 and fingers are $15 apiece. What a killing the body-snatchers would have made. But that was in the bad old days . . .

In 2004 the new owner of a New York funeral parlour was shocked to find that it had a concealed operating theatre and much of its income had come from tissue-transplant companies. The previous owners had treated the dead as commodities to be exploited. The 'operating theatre' had been a cutting room for harvesting human organs. Instead of embalming clients' loved ones they had pillaged their interiors. Stolen bones were replaced with plastic piping and the space left by filched organs was packed with cloth and the morticians' discarded surgical gloves. Then the bodies were sewn up and returned to the family for burial. One of the victims was the revered broadcaster Alistair Cooke. He was ninety-five and his bones were riddled with cancer. The morticians are thought to have made $4.7 million from their sideline.

To make the organs more marketable they had faked the paperwork: a 104-year-old woman was listed as dying aged seventy, those who had nasty diseases that should have precluded their tissues from being used for transplants were said to have succumbed to heart failure. The organs were dispatched all over the world by unwitting tissue-supply companies. Forty people in Britain received transplants from this source. All the known recipients of these dubious organs were tested for HIV, hepatitis C and syphilis, but some diseases don't show themselves for years.

The case is not unique. A doctor in Denver uncovered an identical scam perpetrated by a funeral director who handed relatives any old ashes with one hand while sawing the deceased into useful bits with the other. The total number of bodies desecrated by these two firms was well over a thousand.

Even living patients may be harvested for profit. Human cells that grow readily in culture are useful for studying cancer. In 1990 the California Supreme Court ruled that patients do not own tissue removed from their body. Doctors are entitled to exploit and even patent these cell lines. They are worth millions of dollars and the only person who doesn't profit is the donor.

Like many anatomists, the Hunter brothers collected specimens of organs and bones in their museum. Eventually it contained 13,500 specimens. The poet Southey paints a lurid picture: 'I have made candles of infant's fat . . . I have bottled babies unborn, and dried hearts and livers from rifled graves.' Medical museums were for educational purposes, but anatomists were also fascinated by anatomical oddities and abnormalities and were ruthless collectors. At Guy's Hospital a patient with a curiously enlarged head died. While a mock funeral took place for the unsuspecting relatives, his skeleton was on the slab being prepared for the hospital museum. When Charles Byrne, the nearly eight-feet-tall 'Irish giant', was dying, he made the undertakers promise to sink him in the sea in a lead coffin to avoid the anatomists. But John Hunter offered them £500 (approximately £30,000 in today's money) and Byrne's reluctant skeleton now has pride of place in the Hunterian Museum at the Royal College of Surgeons.

When he became more successful John moved into a new house in Leicester Square – in fact, two houses that he linked together. This gave ample space for his museum and a dissecting room in the attic. Few families can have had quite so many skeletons in their cupboards. Only once was an over-ripe cheese mistaken for a parcel of anatomical over-ripeness. It was a Jekyll-and-Hyde house, for while his cultured wife held soirées in the drawing room to entertain Haydn, dead bodies were being sneaked in by the rear door and dragged up the back staircase. I recall that Stevenson's Dr Jekyll bought his London house from 'the heirs of a celebrated surgeon'.

John Hunter was the most accomplished and innovative surgeon of his day. William Hazlitt described how he 'set about cutting up a carcass of a whale with the same greatness of gusto that Michelangelo would have hewn a block of marble'. Unlike most of his contemporaries he never favoured theory over experience and firmly believed in autopsy (literally, 'to see with one's own eyes').

Surgeons routinely amputated infected or damaged parts of the body, but John's superior skill and perception led to new and better treatments. Aneurysm is a fatal condition which arises when the walls of a blood vessel weaken so that the vessel inflates with blood and can burst. Hunter operated on a patient with a large aneurysm at the back of his leg. By tying off the affected length of the artery, in the belief that the blood would naturally find an alternative route through adjacent blood vessels, he saved the man's leg. It wasn't a wild guess; he had previously done similar surgery on a dog and a deer to see if it would work. When a year later the man died of a fever unrelated to the operation, Hunter bought his body and re-examined the blood vessels in the leg to confirm that his conjecture was correct. By this time his 'bypass' technique was becoming the standard procedure for aneurysms of the leg in all the best hospitals of Europe.

The Hunter brothers made significant advances in medical knowledge. Sometimes their joint projects were William's idea, as when they revealed the extent and function of the lymphatic system, but John did all the dissections and experiments, and it was he who recognised that the lymph system was implicated in a certain type of cancer.

It was also William's plan to reveal all the stages of foetal development inside the mother's womb. Unfortunately, there was a dearth of pregnant women to dissect because felons 'with child' were not hanged. Even if a lass wasn't pregnant when she was arrested, she made sure she was before being sentenced. It was over twenty years before the work was completed. William published it as a huge atlas in 'elephant' folio with every stage of gestation brilliantly illustrated by the Dutch artist Jan van Rymsdyk. It was one of the finest anatomical books ever produced. Although John had done all the work, William gave only a brief acknowledgement that his brother had assisted in 'most of the dissections' – and the artist was not mentioned at all. The text revealed for the first time that the placental blood circulation was independent of the mother's, something that had been discovered by John and a colleague. William took the credit, creating a permanent rift between the brothers.

John published under his own name and again used van Rymsdyk to illustrate his brilliant treatise on human teeth, which introduced the terms 'incisors' and 'molars'. He also recognised that dental plaque was connected with tooth decay and recommended its removal by daily brushing.

John Hunter's experimental methods were the mark of his genius. He once said to his pupil Edward Jenner, who went on to develop vaccination against smallpox, 'I think your solution is just, but why think, why not try the experiment?' It seems obvious now, but this was at the dawn of the age of experimentation.

Hunter investigated the possibilities of artificial insemination and instructed a couple, for whom normal copulation was impossible, how to get pregnant – and the woman did. He also pioneered tissue transplantation by successfully moving organs from one animal to another. He erroneously thought he had got a human tooth to bond into a cockerel's tissue. In those days dentures were manufactured from elephant's tusks. Not surprisingly, they looked better on the elephant. Human teeth were clearly more 'natural' and grave robbers had a lucrative sideline in extracting teeth no longer required by their original owner. Hunter had heard of a better idea. He pulled sound front teeth from paid volunteers and immediately implanted them into the vacant mouths of rich dowagers. A poor girl who would later become Lady Hamilton made a good career choice when at the last moment she changed her mind and didn't donate her smile. None of the transplanted teeth became permanent fixtures, although some lasted for six years and in one case reputedly for twelve years. Hunter inadvertently stimulated a craze for dental implantation, but even false teeth can bite back. Enthusiasm for the implantation of human teeth waned when a woman caught syphilis from her new choppers.

Hunter never shunned controversy. His museum was arranged to demonstrate that 'Every property in man is similar to some property . . . in another animal' and that simian and human skulls fell into a graded series that led to man. These were not popular notions at the time, seventy years before the publication of Darwin's *The Origin of Species*. Equally blasphemous was his assertion that Adam and Eve were indisputably black. It is now accepted that the first humans did indeed arise in Africa.

He soundly refuted the widespread belief that masturbation caused impotency. His logic was persuasive: impotency was rare but masturbation was exceedingly common, so it was unlikely that one led to the other. In his publication declaring that masturbation was not harmful, the embarrassed editor added a footnote to say that it *was*. In certain circumstances it *can* have serious consequences, as in Indonesia where the penalty for masturbation is decapitation. Recent research indicates that 'self-dating', as it's now called, is beneficial. The more men ejaculate earlier in life, the lower their chances of developing prostate cancer later on.

John Hunter was aware that many sexual problems might be psychosomatic, for 'the mind is subject to a thousand caprices, which affect the action of those parts'. When a patient complained that he was a failure in bed, Hunter instructed him to sleep close to his partner for a week without touching her. Seven days later the problem was solved.

Although Hunter had unruly hair and an unruly manner, his reputation was such that the great and the good (and the not so good) flocked to his door. He attended the economist Adam Smith, Benjamin Franklin, Lord Byron, and became 'Surgeon-extraordinary' to King George III.

In 1767, aged thirty-nine, he was elected a Fellow of the Royal Society. That same year he took his studies of venereal disease to a new and dangerous level. All eighteenth-century doctors were familiar with *morbus venereus* (the sickness of Venus) as its victims made up about a quarter of their business. It was well understood that these infections were sexually transmitted. London was a bustling commercial centre for tea, sugar, spices and the most spicy commodity of all – sex. There was one prostitute for every twenty-seven men. Harris's *Man of Pleasure's Kalendar* guided clients around the 'Covent Garden Ladies'. The guide was explicit in praising the pretty doxies and damning the infected, like poor Miss Young who had 'thrown her contaminated carcass on the town again'. No wonder VD was rife among Hunter's patients. James Boswell had nineteen or more bouts of gonorrhoea without learning his lesson. A distinguished husband and wife both consulted Hunter, the approaches of each unknown to the other. The old adage that 'love passes, but syphilis endures' was confirmed by Mrs Beeton who caught syphilis on her honeymoon.

There were two main venereal conditions, clap (gonorrhoea), and pox (syphilis). Clap was common and caused painful peeing and an unpleasant discharge. Although it might lead to complications if you were as keen a devotee as Boswell, it wasn't life-threatening. Pox, on the other hand, was a far more virulent and insidious beast. Initially there would be a lump on the penis and the lymph glands would swell. Within a few weeks all might seem well, but a month or two afterwards wartlike growths would appear, accompanied by patchy hair loss and fever, which could recur repeatedly. Syphilis can remain dormant for years and then reappear, but progressively over time the skin and bones become ulcerated. One patient's penis was so ulcerated that he pee'd over his shoulder. Internally the organs – including the brain – are in meltdown.

Hunter felt sure that clap was a self-limiting disease that could clear up even if left untreated, so he ran clinical trials on patients suffering from it. He gave half of them the usual remedy and the others got pills made from rolled-up bread. In time they all recovered. He also hypothesised that two diseases couldn't occupy one body at the same time, and therefore clap and pox must be different phases of the same disease – clap was the localised infection, which later spread through the body to become pox.

The way to settle the matter was clearly by an experiment. Hunter's idea was to infect someone with the clap and then wait for the symptoms of the pox to appear, thus supporting his hypothesis, or fail to appear, indicating that he was wrong. Clearly, the only volunteer whom he could guarantee wasn't infected with either disease and whose genitalia were close at hand for daily inspection was himself.

So Hunter carefully transferred some of what Boswell called 'the loathsome matter' from a patient with clap into incisions that he had cut on his own penis. Imagine the satisfaction he felt when a few weeks later the characteristic nodule of pox called chancre appeared on his penis. It was later designated 'Hunterian chancre'.

What John hadn't considered was that the patient he had used for the inoculum might be suffering from *both* clap *and* pox. He had inadvertently given himself syphilis which, if not stopped early on, leads inexorably to the corrosion of the nose, blindness, paralysis, insanity and death. You can always rely on a rational man to do something completely irrational.

To cure his pox Hunter repeatedly swilled his mouth with 'corrosive sublimate' and toxic mercury. These substances give mouth ulcers, loosen the teeth and produce pints of black saliva. Some hospitals had 'salivating wards' where one could dribble in private. He later stated 'I knocked down the disease with mercury', suggesting that the treatment was successful. He used his experience in his lectures to students, making it clear that he had caused himself to develop a syphilitic chancre. He also wrote an illustrated treatise describing sexually transmitted diseases that was so graphic it even put Boswell off sex for a week, although he went on to incur the symptoms twice more.

For his cardiac problems Hunter tried all manner of poisons before resorting to 'Madeira, brandy and other warm things' – the efficacy of which I can confirm from my own experiments.

When researching into sexually transmitted diseases, doctors have never baulked at experimenting on others. In the infamous Tuskegee project in Alabama, poor black sharecroppers with syphilis were monitored by doctors for forty years to see if the symptoms progressed in the same way as they did in white men. Syphilis was never mentioned; the men were told that they had 'bad blood'. Although they had to endure invasive tests, not one of them was given any treatment for the disease. The doctors were merely observers. None of them felt any guilt since they hadn't actually given the men syphilis – they'd already had it. Many of the 'patients' went on to suffer horrendous symptoms as the disease progressed.

The study was funded by the United States Public Health Service and was known to the medical community and local politicians. It continued until 1972, when a journalist broke the story nationwide, but it was not until 1997 that President Clinton apologised to the surviving human guinea pigs.

In Hunter's day many surgeons considered that they had no investment in keeping non-paying patients alive. Sir Astley Cooper wrote: 'The patients in our hospital, from whom practical knowledge is to be derived are . . . just as much within the surgeon's power, as dead bodies are at the disposal of Parliament.' John Hunter experimented on patients, but did not disdainfully consider them as mere guinea pigs, 'nor,' he said, 'do I go further than . . . I would have performed on myself were I in the same situation'. He taught his students that: 'No surgeon should approach the victim of this operation without a sacred dread and reluctance.'

Only John Hunter had the courage to experiment on himself. But where he led many others would follow.

A watchman nabs a bodysnatcher while William Hunter exits at speed.

Sniff It and See

'Should my leg be cut off, I would never be chloroformed. I would never want to abdicate myself.' – Honoré de Balzac

John Hunter made it clear that surgery was the last resort. This had always been the patients' view. Surgery was calculated violence. To lie on the operating table was a guarantee of agony with a sincere promise of death. The patient had an unfortunate habit of screaming, which was distracting for the surgeon, and of writhing in agony, which made accurate knife-work almost impossible. What if the patient were drugged into a stupor?

A thirteenth-century physician administered a lively cocktail of opium, henbane, hemlock and mandragora (a relative of deadly nightshade). He claimed that it produced 'a sleep so profound that the patient may be cut and feel nothing as though he were dead'. The patient wasn't fooling – he probably *was* dead. More practical medics bled or choked the patients into unconsciousness or put a wooden bowl on their head and whacked them with a mallet.

By the late eighteenth century the medical profession had learned to cope with the agony of surgery. The rules were simple:

1. Site the operating room out of earshot of other patients.
2. Be excessively solicitous over the distress caused to the poor surgeon.
3. Strap the patient down securely.
4. Have the patient bite down on the surgeon's walking stick.
5. Slice and saw at speed.

Surgeons had to be strong, practical men. John Hunter developed his technique with saws in a timber yard. His brother William called surgeons 'savages armed with knives'. But at least they were quick. William Cheselden, a protégé of Hunter, extracted bladder stones in less than a minute and Robert Liston, the great British surgeon, amputated a leg in twenty-eight seconds. He loved the idea of the operating *theatre* and would commence each performance by announcing to the audience, 'Time me, gentlemen. Time me.' In striving to break his record he not only detached the patient's leg but also one of his testicles, along with two fingers from Liston's assistant.

The author Fanny Burney described the horror of a mastectomy. 'When the dreadful steel was plunged into my breast – cutting through veins – arteries – flesh – nerves . . . I began a scream that lasted unremittingly during the whole time of the incision . . . so excruciating was the agony . . . all description would be baffled . . . I felt the knife rackling against my breastbone – scraping it.'

Operations were no less agonising in Japan until a surgeon called Seishu Hanaoka did something about it. He had learned some surgical techniques from European books, but for the control of pain he turned to Chinese medicine. After twenty years of experimentation on animals he thought he had the right mix of plant extracts that dulled pain without dangerous side effects. He confidently tried it on his wife. She went blind.

Undeterred, he continued his research and eventually arrived at a concoction containing compounds that we now know to be sedatives, analgesics and muscle relaxants. In 1804 he painlessly removed a tumour for a woman with breast cancer and went on to perform 150 pain-free operations. Unfortunately Japan's self-imposed isolation from the rest of the world meant that Seishu Hanaoka's anaesthetic remained a secret.

In Europe some medics were keen to inhale newly discovered gases to see if they had any therapeutic properties. Someone who was sure that they did was Dr Thomas Beddoes. He was somewhat unconventional and conveyed a cow to invalids' chambers so that they might inhale the animal's 'restorative breath', but succeeded only in ruining their bedroom carpets. He established the Medical Pneumatic Institution in the spa town of Bristol. It advertised gas-therapy 'cures' for everything from venereal disease to paralysis by offering snorts of oxygen, carbon dioxide and even what was to become the suicide's favourite, carbon monoxide, because it brought colour to the cheeks.

The Institution's superintendent of research was Humphry Davy. He is best remembered today as the inventor of the miners' safety lamp, but in his lifetime he was enormously famous. He was a gifted communicator and sold science to industrialists as a tool for the production of wealth. By doing so he changed the world for ever.

When Davy came to the Pneumatic Institution he was only twenty-one, but he soon determined that the benefits of the gases were vastly exaggerated. He tested many vapours on himself. After inhaling carbon monoxide he was 'sinking into annihilation, and had just power enough to drop the mouthpiece from my unclosed lips . . . There is every reason to believe, that if I had taken four or five inhalations instead of three, they would have destroyed life immediately.' Breathing pure oxygen immediately afterwards saved his life. Scares didn't curb his enthusiasm. A week later he sniffed a volatile solvent that seared his epiglottis and made him choke. During these ordeals he calmly monitored his pulse rate even when he thought he was dying.

Davy's colleagues were always relieved to see him the next morning. One commented that he risked his life 'as if he had two or three others in reserve, on which he could fall back in case of necessity'.

Following reports that nitrous oxide was a 'principle of contagion' that would be instantly lethal to animals, Davy gave it a try. 'I was aware of the danger of this experiment,' he admitted. Nitrous oxide is an intoxicant and in those days impurities might render even the most innocuous gas lethal. He increased the dose progressively until he was breathing it three or four times a day for a week.

Luckily there were no ill effects, indeed the gas produced a 'highly pleasurable thrilling . . . I lost all connection with external things: I existed in a world of newly connected and newly modified ideas. I theorised, I imagined, I made discoveries.' Davy enjoyed theorising so much that he was soon taking a staggering twenty-five litres a day. He thought the gas might also improve his poetry (it didn't). He gave the gas to his friends, Roget of *Thesaurus* fame and the poets Southey and Coleridge. Southey wrote: 'It made me laugh and tingle in every toe and finger tip. Davy has actually invented a new pleasure . . . I am sure the air in heaven must be this wonder-working gas of delight.' Delight was just what the well-to-do of Bristol were seeking and the number of customers at the Institution increased substantially.

Then serendipity, the researcher's friend, took a hand.

Davy's emerging wisdom teeth caused a painful inflammation of the gums, but when he breathed nitrous oxide the
pain vanished. Never slow to spot an application, he wrote
that as nitrous oxide 'appears capable of destroying physical
pain, it may probably be used with advantage during surgical
operations'. Here was the miracle that every patient craved:
a pain-free operation. Inexplicably, Davy never followed this
up and although many medical students also embraced
nitrous oxide as a bringer of euphoria, no one thought of it
as a practical painkiller for well over forty years.

Nitrous oxide, which Davy called laughing gas, became
a party staple. The naturalist Schoenbein attended a garden
party at which the gas-fuelled antics of guests laid waste
to the flower beds. 'Maybe,' he mused, 'it will be the custom
for us to inhale laughing gas at the end of a dinner party,
instead of drinking champagne, and in that event there
would be no shortage of gas factories.' Nothing has changed.
In July 2007 the BBC news reported concerns that the latest
craze for clubbers was breathing nitrous oxide from
balloons.

Laughing gas also became a fairground favourite. Just as
nowadays hypnotists cajole volunteers onto the stage to
make fools of themselves under the influence, showmen
once extracted amusing antics with a whiff of gas. The volunteer was told to pinch his nose and inhale gas from a bag.
When the bag was withdrawn he sat in a trance 'still holding
his nose. You can imagine how this comical posture sent the

audience into roars of laughter which increased when the intoxicated man leapt smartly from his chair and then made astonishing bounds all over the stage.'

In 1844 'Professor' Colton brought such excitements to the entertainment-starved folk of Hartford, Connecticut. In the audience was Horace Wells, a local dentist. He had developed improved dentures but failed to make the fortune he anticipated, for to fit the false teeth all the remaining rotting stumps and roots had to be extracted. In those days tooth-pulling was an excruciating and bloody business. Toothache was often dulled at the mere sight of a dentist.

During Colton's show one of the volunteers making astonishing bounds cracked his shin. When he returned to his seat, Wells asked about his leg. Although he'd received a nasty gash he felt nothing. Wells immediately grasped that the gas had dulled the pain. He persuaded Colton to bring the apparatus to his house the next day. Colton administered nitrous oxide to Wells himself and a fellow dentist yanked out one of his teeth. Wells rejoiced: 'It is the greatest discovery ever made. I didn't feel as much as the prick of a pin.' He could banish pain and bring love at last to dentists.

In the following days Wells pulled teeth painlessly from fifteen patients. He realised that he must demonstrate pain-free extraction to a medical audience and within a month of his discovery he was scheduled to do so at the prestigious Massachusetts General Hospital in Boston. It was too soon. He did not yet understand that unconsciousness did not necessarily mean insensitivity to pain, nor did he realise that different people might respond differently to the same dose of gas. The audience of medical students was restless, and Dr J. C. Warren, the surgeon who introduced him, was clearly sceptical: 'This gentleman,' he announced, 'who pretends he has something which will destroy pain in surgical operations, wants to address you.' The word 'pretends' made Wells nervous. The trial patient, also understandably nervous, accidentally knocked the instruments onto the floor and the audience laughed. Wells hastily anaesthetised the patient and pulled the tooth. Although there were none of the usual screams and struggles, the patient emitted a loud groan that echoed round the auditorium. The audience responded with jeers and cries of 'Humbug!'

Wells had given too little of the gas and the patient was not fully under. It was a disaster; a public humiliation from which Wells never recovered. The rest of his life was spent in regret and recrimination and seeking solace by sniffing nitrous oxide. The gas that was supposed to make his fortune ruined his life.

Ironically, nitrous oxide would become the dental anaesthetic of choice. By 1883 the Poe Chemical Works, owned by Edgar Allan Poe's cousin George, supplied it to 5,000 dentists across the United States, including one enthusiast in Cleveland who ordered 4,000 gallons.

To add insult to Wells's indignation, it was a former colleague who brought anaesthesia to the public's notice.

William Morton was a dentist and an opportunist. Morton's old chemistry tutor, Dr Charles Jackson, suggested that sulphuric ether (derived from mixing sulphuric acid and alcohol) might be a better bet than nitrous oxide.

As the growing temperance movement began to hit sales of liquor, sipping diluted ether gained popularity. Communal ether 'frolics' were commonplace. The patients treated for chronic ether intoxication by a famous London doctor included 'persons of education and refinement . . . mostly women; the men were all doctors'. Morton called ether the 'toy of professors and students', and its widespread social use convinced him that it was safe. Perhaps he was ignorant of the numerous complications that ranged from vomiting to death. Ether vapour was also said to be explosive – not ideal in a world lit by naked flames. So Morton inhaled it and then breathed out into a flame. Luckily there was no explosion. It only ignited when lit by a taper.

Before testing its effects on a human, he decided to try it on something disposable – his wife's pet dog and her goldfish. Both they and his marriage survived, just. He then went out in search of volunteers to be knocked out by the gas and have a tooth yanked to see if it hurt. Not surprisingly there were none, not even for five dollars per head. So Morton shut himself in his office, soaked a cloth in ether and put it over his mouth. With ether it was easy to overdose. A single breath could take in a dangerously large amount of vapour. He collapsed almost immediately, alone with no hope of resuscitation. Fortunately, the cloth fell from his face and about eight minutes later he regained consciousness. When told of his risky experiment, his wife was distraught. The next trial was for his assistant to extract one of Morton's teeth under ether. By chance a 'cracker maker' with severe toothache came to the surgery and willingly took his place. The rotten bicuspid was wrenched out and the patient felt no pain.

The confident Morton arranged a demonstration at the very same hospital where Horace Wells had been humiliated. Morton was late and the operation was due to begin. Dr Warren, who had chaired Wells's disastrous demonstration, was impatient, but Morton anaesthetised the patient quickly and, with the flourish of an actor, announced, 'Your patient is ready, sir.' The surgeon cut into the neck and removed a tumour almost the size a billiard ball. To everyone's amazement the patient was silent and when the operation was finished Dr Warren declared, to loud applause, 'Gentlemen, this is no humbug.' The patient said he had felt no pain; it was merely as if his neck were being 'scraped with a hoe'.

Three weeks later a young woman called Alice became the first to have a leg amputated under anaesthesia. As the surgeon finished the job he roused her and said, 'Are you ready?' She replied, 'Yes, sir.' The surgeon responded with, 'Well, it's done!' and brandished her detached leg. Alice fainted with shock.

The word soon spread and within two months of Morton's demonstration Robert Liston, the lightning leg-lopper in England, used ether for an operation and exclaimed: 'This Yankee dodge beats mesmerism hollow.' Oliver Wendell Holmes coined the word 'anaesthesia' (Greek for 'without feeling') and wrote in an essay: 'Inhale a few whiffs of ether, and we cross over into the unknown world of death with a return ticket.'

Morton was in it for the money or, as he put it, 'personal rights and benefits'. But there was a problem. Ether was as 'free as God's sunshine'. He could not patent a chemical that had been known since 1540 so he tried to sell it under the name of *Letheon* (after Lethe, the stream of oblivion running through Hell). But doctors soon twigged that it was nothing more than good old ether. Morton abandoned dentistry to pursue recognition for having invented anaesthesia, but rival claimants were appearing on all sides. Crawford Long, a doctor in Georgia, had four years earlier successfully excised tumours under ether and had affidavits to prove it, but he had not published and brought his findings to the attention of the medical profession.

Charles Jackson, who had suggested the use of ether to Morton, insisted that he had discovered anaesthesia. However, he was also claiming to have invented gun cotton and was suing Samuel Morse, asserting that he had given him the idea of the electric telegraph. He was staking more claims than an excited gold prospector.

Morton's battle to establish his priority wrecked his health and reading an article in support of Jackson's case almost precipitated a nervous breakdown. He died of a stroke, penniless and as yet unrecognised. But Jackson did not triumph. He became an alcoholic and was found at Morton's graveside screaming at his dead rival. He died in an asylum.

Morton's former partner, Horace Wells, fared no better. He was incensed that amid all the rival claims for ether his earlier work with nitrous oxide had been forgotten. He was now a chloroform salesman busily sniffing away his profits. In an apparent attempt to rid the streets of prostitutes he threw acid at two women and was arrested. In prison, under the influence of chloroform, he slashed an artery and bled to death. He was only thirty-three years old. His estranged wife accused Morton of having stolen Wells's discovery and driven him to madness and suicide. On the day she heard of his death a letter arrived stating that the Medical Society of Paris had recognised him as the discoverer of anaesthesia. Now his only recognition is on his tombstone, and a plaque on the wall of the Burger King restaurant in his home town of Hartford.

Today in the USA alone five to ten million patients per year receive nitrous oxide in the cocktail of anaesthetics given during operations. But, at least for major surgery, its days may be numbered. Recent research has shown that replacing it in the mix with oxygen greatly reduces the frequency of life-threatening complications.

Although ether also became widely used in surgery, it had an awful smell, it irritated the lungs and before patients were fully under they tended to thrash about. At Edinburgh University the young Professor of Midwifery James Simpson, a protégé of Lightning Liston, began to use ether to relieve the pain of women during childbirth. But he felt there must be a better, as yet undiscovered, anaesthetic. So he began sniffing anything that had a vapour. His solvent soirées with medical friends, and relations too, culminated not in passing the port but in breathing whatever solutions he had come across since the last party. This was exceedingly dangerous as he often had little knowledge of the properties or toxicity of the substances concerned. Among the solutions tested were acetone, now familiar as nail-polish remover, ethyl nitrate, a constituent in rocket fuel, and benzene, a poison and potent carcinogen. Such horrors were rejected largely because they smelled bad, caused headaches or irritated the lungs. With less luck, they might easily have been marked down for having killed the entire gathering. During one sniffing session in 1847 they tried sweet-smelling chloroform. Simpson's first impression was: 'This is far stronger and better than ether.' He then realised that he was lying on the floor. Indeed, 'We were all under the mahogany in a trice.' They tried it again and down they went. Even his wife's young niece, who had only the slightest whiff, exclaimed, 'Oh, I'm an angel.'

Chloroform had been discovered by an inadvertent self-experimenter, Samuel Guthrie. He was an inventor fascinated by explosives and determined to improve on gunpowder. By his own admission his experiments caused hundreds of accidental explosions, one so violent that it blew the roof off his workshop and demolished walls. Guthrie frequently had to run for his life. There was hardly a bit of his body that hadn't been partially barbecued at one time or another.

In 1831 he began to investigate the commercial potential of chloric ether, a supposed stimulant, which is now used as a pesticide. When he brewed up the constituents he made not chloric ether, but an alcoholic solution of chloroform – although he didn't know it. It was an 'intensively sweet and aromatic' tipple when diluted with water so he sold it locally as Guthrie's Sweet Whiskey. It became so popular that even respectable old ladies were passing out at the roadside. He little guessed that its ability to bring oblivion was its most commercial property.

Sixteen years later, and only four days after his first sniff of chloroform, James Simpson administered it to a pregnant woman who had on a previous pregnancy endured a three-day labour and had lost the baby. When she came to, the patient couldn't believe she had given birth. She christened her new daughter Anaesthesia. Within weeks chloroform was being used in all operations at Edinburgh Royal Infirmary.

Although most surgeons embraced anaesthetics, a vociferous minority insisted that pain was good for you. One wrote to a medical journal that anaesthesia was a 'decoy by which the credulous may be induced to give up their senses as well as their cash'. Another wrote that: 'Knife and pain in surgery are words which are inseparable in the minds of patients and this necessary association must be conceded.' The strongest objections were against giving anaesthetics to women during labour. 'Pain,' another male doctor declared, 'was the mother's safety, its absence her destruction.' Charles Meigs, Professor of Midwifery at Jefferson Medical College in Philadelphia, assured women that labour pains were 'a most desirable salutary and conservative manifestation of the life force' and that there was a 'needful and useful connection of the pain and the powers of parturition, the inconveniences of which are really less considerable than has by some been supposed'. But then, I don't think he had ever given birth himself. He seems not to have held a high opinion of women, having pronounced that their heads were 'almost too small for intellect and just big enough for love'.

It didn't matter what the doubters said; expectant women began to demand chloroform. They were encouraged when notables gave it their blessing. After his wife Catherine had a pain-free delivery, Charles Dickens wrote that chloroform was 'as safe in its administration as it is miraculous in its effects'. When Queen Victoria opted for chloroform for her last two births and called it 'delightful beyond measure', its popularity was assured.

It was nowhere more popular than at parties in Professor

Simpson's Edinburgh residence where 'instead of music and dancing . . . every guest was treated to a trip to the realms of insensibility'. The widow of a local physician later recalled that in her youth 'the Professor used to try his experiments with chloroform on us girls. Our mother feared nothing, and was only too delighted to sacrifice, if unavoidable, a daughter or two to science.'

Simpson became the apostle of chloroform and even enlisted God as the first anaesthetist. When extracting a rib to create woman, He 'caused a deep sleep to fall upon Adam, and he slept: and he took one of his ribs and closed up the flesh'.

Simpson extolled chloroform's virtues at every opportunity and was blind to its faults. When a young girl died within two minutes of chloroform being administered for an operation on an ingrowing toenail, Simpson dismissed suggestions that the anaesthetic was to blame. He *knew* it was safe – after all, he had tested it on himself.

But the death rate mounted. Eventually, a review of over 800,000 operations under anaesthetic revealed that the death rate from chloroform was four and a half times higher than for ether. Many of these deaths, even those of vigorous young people, were almost instantaneous, as if the patient had been shot through the heart. It was many years before medics established that with chloroform there was a fine line between anaesthetisation and a fatal overdose that caused heart failure.

Though ether and nitrous oxide gradually came back into favour, Simpson never failed to champion chloroform, even though it probably killed one in ten of his patients. He also continued to jeopardise his health with further self-experimentation. 'I am ill and quite undone,' he wrote, 'from breathing and inhaling some vapours I was experimenting upon last night, with a view to obtaining other therapeutic agents.' His servant found him unconscious on the floor and feared for his life. He wondered why Simpson took such risks, for he will 'never find anything better than "chlory"'. Obsession is the hallmark of many self-experimenters.

It is estimated that over 100,000 people died from the medical use of chloroform. Simpson used it for the last two years of his life to control his angina. He died in 1870, worn out by overwork and self-experimentation. He was lucky to make it to the age of fifty-nine.

Anaesthesia revolutionised surgery. It prevented the patient having to undergo a terrible ordeal and succumbing to what Joseph Lister called 'mortal shock'. There were those who believed that the patient suffered just as much pain but forgot about it on awakening. Dr Warren at Massachusetts General, the first surgeon to operate with an anaesthetic, had a singular idea of what the patient felt: 'Who could have imagined that drawing a knife over the delicate skin of the face, might produce a sensation of unmixed delight? That the turning and twisting of instruments in the most sensitive bladder, might be accompanied by a delightful dream?'

Longer and more invasive operations could now be attempted, but it would be forty years before the first local anaesthetics came along. Cocaine was first promoted as a treatment for morphine addiction. It was an early ingredient of Coca-Cola, which was marketed as a remedy for depression and hysteria. No doubt it also attracted many new drinkers. It is amazing how blasé we once were about powerful drugs. During the First World War the famous London store Harrods offered 'a gift box for our friends abroad'. It contained vials of morphine and heroin with a syringe.

The young Sigmund Freud, with his self-styled 'explorer's temperament', began taking cocaine to test its effects as a stimulant and an aphrodisiac. His trials led him to extol its virtues and to assure his readers unwisely that 'even repeated doses produce no compulsive desire to use the stimulant further'. It was one of the greatest blunders of his career. He went on to turn several of his patients and himself into addicts.

He did, however, notice its tendency to numb the tongue and mentioned this to an ophthalmologist. If Freud had devoted his time to following up this observation, the world might have been spared countless hours of psychoanalysis and outbreaks of the Oedipus complex. It is amazing how many medical men also noted the numbing effect of cocaine but failed to appreciate its possible significance. Freud's ophthalmologist, Carl Koller, wondered whether cocaine might also numb the eye to allow surgery. So he and a colleague drizzled a solution of cocaine into their eyes and then jabbed the cornea with a pin and felt nothing but pressure. Koller established cocaine as the ideal anaesthetic for eye surgery and became known as 'Coca Koller'.

This may have encouraged two New York surgeons, Richard Hall and William Halsted, to inject each others' limbs and gums to produce localised areas of insensitivity to pain. Both became addicted to cocaine. Halsted took morphine to 'cure' his dependency on cocaine and became addicted to morphine for the rest of his life.

In 1886 the lumbar puncture (spinal tap) was first used to collect spinal fluid from a living patient. A German surgeon, August Bier, realised that this technique might allow cocaine to be introduced into the spinal cord to block the nerves serving muscles below the point of injection. So his assistant, August Hildebrandt, pushed a hollow needle through the membranes of Bier's spinal cord and into the fluid-filled cavity. While he fumbled to attach the needle to a syringe of the wrong size, Bier's spinal fluid was leaking out onto the floor. The assistant plugged the leak and became the next volunteer. His injection went well. During the following half an hour Bier enthusiastically tickled the soles of Hildebrandt's feet with a feather, pinched his skin with hooked forceps, jabbed his thigh to the bone with a surgical lance, plucked hair from his pubes, stubbed out a lighted cigar on his skin, whacked his shins with a heavy hammer and for a finale violently squashed and yanked his testicles. A thorough man, Bier left no place untortured. Fortunately the whole lower half of Hildebrandt's body was insensitive to pain – until the effects of the cocaine wore off. This experiment transformed surgery on the lower body. Bier went down in medical history and Hildebrandt is remembered as the man whose testicles he tugged.

For medical and dental purposes cocaine has been replaced by less addictive synthetic substances that have been developed by self-testing newly synthesised drugs. When not at the dentist, however, we prefer the addictive version. Around eighty per cent of all banknotes in circulation in Britain are contaminated with cocaine or heroin, rising to ninety-nine per cent in London. More than fifteen million pounds' worth of notes are destroyed each year to protect the non-snorting public.

As no anaesthetics were entirely safe, surgeons continued to operate rapidly rather than expose the patient to long periods of anaesthetisation. What was needed was some way of reducing the large amount of anaesthetic needed to relax the muscles. Perhaps a non-anaesthetic might do the job. The solution came from aboriginal tribes in South America whose hunting arrows and darts were tipped with curare. Its ability to kill almost instantly did not suggest it would be useful for surgery. But even the natives knew that though it was lethal when it penetrated the skin, it was usually harmless when eaten in small quantities. Animal experiments in Europe showed that curare paralysed the breathing muscles but did not still the heart; with artificial respiration the animal could recover.

In 1944 the drug company Burroughs Wellcome isolated tubocurarine, the active ingredient of curare. Frederick Prescott, its head of clinical research, was not averse to testing new drugs on himself. He had taken a combination of morphine and methamphetamine ('speed') to test a theory that they might help to control blood pressure. In fact Prescott's blood pressure soared and he had to be hospitalised.

Undeterred, he volunteered to see whether curare might have some benefit for surgery. He also consented to be the human guinea pig. Perhaps he didn't anticipate just how hazardous this might be.

The initial tests gave no hint of the ordeal to come. To see if curare could reduce pain, large strips of sticking plaster were ripped from Prescott's body. Curare was certainly no anaesthetic.

Then the trials got serious. To ensure a cosily realistic atmosphere Prescott lay on the table in an operating theatre. He was injected with tubocurarine, the medical equivalent of the poison dart. At intervals over a period of two weeks, with an anaesthetist and doctors in attendance, the dosage was progressively increased. Within two minutes of the final experiment commencing, Prescott's face, neck and all his limbs were totally paralysed. A minute later his breathing muscles were also paralysed – and no one noticed. It was a nightmare worthy of Edgar Allan Poe. He could hear his colleagues chatting but was unable to move a finger or even an eyelid. He was helpless, terrified and sinking into unconsciousness.

Although the rest of the team were monitoring Prescott's blood pressure and his super-fast pulse, they were unaware of his terror. He had not turned blue because they were pushing air into him by squeezing a rubber bladder. When at last they decided they had collected enough data, they injected him with an antidote to reverse the effects of the curare. It took a further seven agonisingly long minutes of artificial respiration before Prescott could breathe for himself. It was over four hours before he could see properly. Other ill effects lasted for days.

The experiments had been carefully planned, but no one had established a system by which the human guinea pig could signal if he was in distress. In more recent experiments the volunteer has a tourniquet on one arm to isolate it from the curare in his system. Thus he can communicate with the team through pre-arranged finger signals.

Despite his ordeal, Prescott volunteered for another experiment that lasted for forty-five minutes. As a consequence compounds like curare are now used universally as an adjunct to anaesthetics to incapacitate patients undergoing surgery.

Four of the pioneers of anaesthesia became addicted to the drugs they were testing. Many of them died prematurely as disappointed men, having failed to achieve the recognition they thought they deserved. In contrast, Frederick Prescott was a modest chap who expected no praise outside the circle of his colleagues. Most of his family only learned of his risky adventures when they read the details in his obituaries many years later.

Anaesthesia has come a long way since the whack on the head with a mallet, but it is still a tricky business. In the United States no less than a hundred patients a day are said to regain consciousness while under the knife. Carol Weihrer has recently recommended the use of a wakefulness monitor in operating theatres after she 'awoke' during an eye-removal operation. 'I didn't feel any pain,' she said, 'but I felt tremendous pulling. It takes a lot of torque to get an eye out.'

Trials and Tribulations

'We often make people pay dearly for what we give them' —
Comtesse Diane

Even with anaesthetics, operations remain a daunting prospect so it is comforting to believe that there is a gentler treatment for our illness. We put great faith in 'cures'. Patients who depart from the consulting room without a prescription may feel that they have been cheated. Our faith in drugs is so powerful that even if given a pill with no curative powers whatsoever, many people feel better. There has always been the whiff of magic in medicine.

The ancients had an appetite for noxious nostrums for diseases, and love philtres for unrequited desires. Herbalism was the midwife of both botany and medicine. For two hundred years virtually all botanists were also physicians or apothecaries, and herbals were the pharmacopoeias of their day. John Gerard's *Herball*, published in 1597, became the most famous botanical book ever printed.

Unfortunately, herbalists were shackled to the 'doctrine of signatures', a belief that the nature of its healing powers was stamped on every plant. A resemblance to a part of the human body indicated the organ that it treated, hence lungwort, bladderwort, toothwort, and anything with a suggestive protuberance was of course an aphrodisiac. To the active imagination mandrake root was shaped like a naked manikin and was reputed to scream when uprooted from the ground. It could cause paralysis and madness, so it was used in love potions.

Because plants are treasure houses of drugs, a few of the herbalists' concoctions actually worked: Jesuit's Bark (quinine) from the *Cinchona* tree was indeed a treatment for ague (malaria), digitalis from foxglove leaves *is* a heart stimulant, senna pods *are* a useful purge, and centuries later salicylic acid from willow bark would be marketed as Aspirin.

In 1733 *Poor Richard's Almanac* contained sage advice: 'He's the best physician that knows the worthlessness of the most medicines.' Although they promised much, the majority of the herbalist's remedies were useless or even dangerous. A buttercup ensured that he who takes it 'leaves this life laughing', but the herbal fails to make clear whether without the help of the buttercup he might not have left at all.

When fewer surgeons were part-time barbers and doctoring became a semi-respectable trade, 'trained' surgeons and physicians emerged from universities. In the eighteenth century a medical degree could be bought for twenty pounds even from some of the most prestigious institutions in Scotland without having to attend a single lecture. The standard of the student intake, even as late as 1869, may be judged from a statement by the Dean of Harvard Medical School who revealed that they set no written examinations because the 'majority of students cannot write well enough'. A graduate who *had* suffered years of examinations and dissections qualified 'never having dressed a wound, given an injection, seen a birth or attended at the bedside in a professional capacity'. Even the most experienced practitioners were powerless in the face of almost all the great killer diseases until the dawn of the twentieth century.

Many people did not survive infancy, because of the ravages of incurable chickenpox, diphtheria, dysentery and measles. Legions of adults were cut down by meningitis, tuberculosis and pneumonia; sex came seasoned with syphilis, and giving birth included the horrors of childbed fever brought by surgeons coming to the delivery room straight from the mortuary.

Confronted by such foes, doctors fell back on Voltaire's advice: 'The art of medicine consists in amusing the patient while nature effects the cure' − or until he dies. The helpless physician could only indulge in 'heroic cupping and vigorous blooding' and refining his bedside manner. When prescribing drugs he plumped for the old standbys: poisonous antimony for fever, with side effects such as cardio-vascular failure and sudden death; generous doses of toxic mercury against venereal disease with the added benefits of loss of teeth and floods of bloody diarrhoea followed by circulatory and kidney failure; opium for pain relief and a lifetime without relief from addiction. A century later Mark Twain could still claim that 'a natural death is where you die without a doctor'.

The quality of doctoring may be judged by that of the royal physician who treated the madness of King George with leeches to the legs, a stream of emetics and by blistering the monarch's shaved scalp to 'extract the poisonous matter from his brain'. No wonder an eighteenth-century cynic wrote: 'It is often ask'd, what Disease a Man died of ... but properly speaking, the Question should be ... what Doctor he died of.' It is likely that the best-selling powders of an Oxford-trained doctor accelerated the demise of numerous patients including Laurence Sterne and Oliver Goldsmith.

Physicians were also inordinately expensive. Even the wealthy baulked at their bills. George III accused one of his physicians of belonging to a profession that 'I most heartily detest'. The medic protested that Jesus healed the sick. 'Yes,' said the King, 'but not at £700 [a year] for it.' Later the King's widow spent so much on doctor's bills that she had to sell most of the furniture from her home at Frogmore.

Untrained quacks on the other hand were cheap. In eighteenth-century England anyone could patent a medicine and, as the ingredients were not regulated, the content could be adulterated at random. To describe the status of these dispensers a contemporary invented the phrase 'licensed to kill'. The satirical cartoonist William Hogarth called quacks 'The Company of Undertakers'.

Daniel Defoe described how: 'Corners of Streets were plaster'd over with Doctors' Bills and Papers of ignorant fellows; quacking and tampering in Physick, and inviting the People to come to them for Remedies.' All the treatments were of course 'Infallible'. Medicine was big business and some quacks made a fortune. Careless experiments in the bedroom led to a brisk demand for cures for the clap. In 1750 three out of every four doctors relied on venereal infections for their business.

Mountebanks were distinguished from other practitioners by the shameless promotion of their products. If their handbills were to be believed, their miracle potions were more than a match for any disease:

> Here take my bills,
> I cure all ills,
> The Cramp, the Stitch,
> The Gout, the Itch,
> The Squirt, the Stone, the Pox,
> The Mulligrubs,
> The Bonny Scrubs,
> And all Pandora's box.

At a penny a pill, how could a wretch with mulligrubs resist? An impoverished population riddled with disease was vulnerable to shysters who 'pretend to perform matters beyond reason'. *Winter's Elixir Vitae* was advertised as having 'revived great numbers of people supposed to be dead'. It was ever thus: the 'sick trade' is lucrative and misfortune is ripe for exploitation.

There were also so-called 'piss prophets' who could diagnose any illness with a glance at the patient's urine. When 'Doctor' Myerbach was handed a flask of cow's pee he instantly identified 'too great a pleasure in women'.

Ben Jonson condemned the itinerant quack as 'a turdy-facy, nasty-paty, lousy fartical rogue' more interested in lucre than the welfare of a patient he would never meet again. Most quacks were showmen as well as salesmen. They would first draw a crowd and then a tooth or two, with loud music to drown the cries, and finally hawk remedies such as *Scot's Pills* – 'excellent for hard drinking', *Rose's Balsamic Elixir* for curing 'Frenchify'd (i.e. syphilitic) patients' or *Horseballs* for coughs.

While quacks charmed the punters with their slick, persuasive patter, educated physicians bamboozled them with impressive jargon. Fielding lampooned them with the doctor in *Tom Jones*: the patient, he opined 'had received a violent contusion in his tibia by which the exterior cutis was lacerated, so there was profuse sanguinary discharge . . . Some febrile symptoms intervening at the same time (for the pulse

was exuberant and indicated some phlebotomy), I appre-
hended an immediate mortification.'

Even reputable physicians learned largely from experience,
and some quacks became more adept at tooth-pulling and
bone-setting than most qualified doctors. Sir Hans Sloane,
the President of the Royal College of Physicians, chose 'Crazy
Sally' Mapp, the 'Epsom bone-setter' to treat his niece's
chronic back problem rather than risk the ministrations of
his fellow doctors.

In Georgian England people could pick and mix what
treatment they used. They often turned to self-dosing in the
belief that taking control of their treatment was safer. Self-
dosing became known as 'quacking yourself'. Horace
Walpole said that his father, the statesman Sir Robert
Walpole, had 'quacked his life away'.

To guide the self-experimenter there were do-it-to-
yourself books which the poet Southey thought should be
entitled *Every Man His Own Poisoner*. This was a time when
even the official pharmacopoeias on which doctors relied
endorsed the medicinal virtues of ground woodlice, crab's
eyes, and crushed bedbugs in white wine. They had only
just abandoned recommending unicorn's horn. The self-
help guides and herbals were similar collections of improb-
abilities. *The Poor Man's Medicine Chest* advised stocking up
on such essentials as extract of toxic lead 'for rough but
safe purges', and saltpetre (an ingredient of gunpowder),
perhaps for even less safe purges. Some purges were said

to work fifteen times or more with a single dose.

Medical guidebooks became best-sellers and gave rise to 'schools' of self-treatment. Isaiah Coffin's book stimulated 'Coffinism', a term offering little prospect of a happy outcome. Samuel Thomson's *Physick for Families* went through numerous editions and became a standard health manual for over 150 years.

The first thing was to denigrate the opposition. Thomson devoted an entire section to 'How doctors shorten the lives of their patients'. The book is a mix of good sense and nonsense. His axiom was 'an ounce of prevention is better than a pound of cure'. He rightly rails against doctors treating fever with 'bleeding, blistering, starving, all their refrigeratives, their opium, mercury, arsenic, antimony, nitre, etc. are so many deadly engines combined with the disease, against the constitution and life of the patient.'

He was especially agitated by the practice then (as now) of keeping feverish patients cool to bring their temperature down. Thomson assures the reader that 'No person yet died of fever.' The culprit was 'cold', so the answer was to toss more coal on the fire. After all, he reasoned, the dead are cold so cooling kills. Thomson was an American and the bee in his buckskin was the deadly imbalance between body temperature and the outside temperature, resulting in 'obstructed perspiration', the 'cause of all disorders'. In the 'constant warfare' between cold and heat, cold phlegm nourished worms and internal cooling manifested itself as dropsy, dysentery, canker, tuberculosis, pleurisy, even birth pangs.

Thomson also provided handy hints on what to do if you were spiked in the eye with a pitchfork or bitten by a mad rat. He describes how he cured a stinking black gangrenous foot with nothing more than a meal poultice.

Such manuals allowed people to self-diagnose, which added to the dangers of self-medicating. It was easy to get hold of virtually any remedy. Provincial papers were kept afloat by the revenue from adverts for proprietary medicines such as *Daffy's Elixir* and *Storey's Worm Cakes*. One powdered panacea could cure everything from cancer to red hair.

Queen Victoria scoured the papers for the latest remedies. She was a renowned imaginary invalid. Physicians had always cultivated the worried well and did little to dispel unfounded concerns in their wealthier patients. Self-diagnosis nurtured morbid health worries in the bourgeoisie to such a degree that hypochondria became known as the 'English malady'. Those in denial boasted that hypochondria was the only disease they didn't have.

Certain ailments became popular until they slipped out of fashion; 'nostalgia' had been commonplace until supplanted by 'the vapours' which in turn gave way to biliousness, with melancholia waiting its turn in the wings. Many of those who had escaped the clammy grasp of the physician fell into an obsessive craving for cures and the embrace of addictive drugs. Queen Victoria was never without *Brown's Chlorodyne*, a heady mixture of chloroform, cannabis and morphine.

Commercial quackery suffered a severe blow at the close of the nineteenth century. With influenza sweeping across Russia and Europe, a company marketed their *Carbolic Smoke Ball*, an inhaler containing powder impregnated with phenol. They boldly offered a reward of £100 to anyone who caught the flu after using their device. Louise Carlill, who unluckily for them was married to a lawyer, used the smoke ball assiduously, caught the flu and sued the firm. The company's defence was that the adverts were 'mere puffery' and only an idiot would believe such extravagant claims. The judge ruled that a vendor making such promises 'must not be surprised if occasionally he is held to his promise'.

Today, thanks to the Web, self-diagnosis is again on the rise and many un-examined people are self-dosing with drugs that have never seen a pharmacy and may even be fakes. Quackery still flourishes. Britons spend £4.5 billion on 'alternative medicine'. In the United States people visit purveyors of untested, unregulated and unlikely remedies more often than they consult conventional physicians. Numerous people still drink urine in the belief that it can cure everything including diabetes and cancer. There may even be an opening for a modern 'piss prophet'.

For centuries the safety of proprietary drugs was effectively tested on the population at large. Most drugs are dangerous if the patient is overdosed, and no one knew what the safe levels were. Prescribing was a case of suck it and see – the patients sucked it and the physician would see whether they died or got better.

Things began to change when medics stepped into the breach to act as human guinea pigs. In 1803 Friedrich Serturner, a young German pharmacist fired with the scientist's 'need to know', set out on a dangerous career of self-experimentation. In spite of fanciful theories, no one knew why even the most useful drugs worked, so Serturner tried to find their active ingredients. He developed a method for systematically analysing the constituents of drugs.

His first success was to show that the principal actions of opium were dependent on a substance he called morphine (from Morpheus, the Greek god of sleep and dreams). He first tested the effects of pure morphine by salting the food of house mice and feral dogs. It put them to sleep – permanently. Undeterred, he tried morphine on himself and his pals to determine the safe dose. They began by ingesting far more than is now considered safe. Soon they became feverish and nauseous with severe stomach cramps. Clearly they had poisoned themselves, perhaps fatally. Drinking pure vinegar induced vomiting and they fell unconscious. Thanks to the vinegar they survived, but the pains lasted for days.

Serturner continued to experiment with morphine and found that when opium failed to dull the pain of toothache a small dose of morphine did the trick. He hoped that 'qualified physicians might soon concern themselves with this matter, because opium is one of our most effective drugs'.

His analytical methods were refined and used to isolate numerous useful compounds related to morphine including adrenaline (a heart stimulant); caffeine (a stimulant used in tonics and painkillers); cocaine (a local anaesthetic); codeine (a sedative and painkiller); ephedrine (used to treat asthma and hay fever) and many more.

Serturner went on to test other drugs and had many nasty moments. He was not alone. In 1819 a Czech medical student called Johannes Purkinje also began to sniff and swallow drugs to test their action. He was aware of the dangers and knew that a self-experimenter should 'exercise care lest he offer himself as a sacrifice by exposing himself to danger'. Nonetheless, he took risks.

Digitalis was widely used as a heart stimulant, but it also blurred the patient's vision. So Purkinje overdosed on digitalis to study what was going on – a dose only a tenth as big had killed laboratory animals. Although his heartbeat became worryingly erratic and his eyesight was disrupted for a fortnight, he was able to make fundamental observations on the nature of vision.

Much of what was known about eyes came from examining cadavers, and the theory of lenses. Purkinje investigated his own vision and in all senses shone a light on the inner workings of the human eye.

He carried out dozens of self-experiments to test the safety of drugs, including belladonna (extracted from *Atropa*, deadly nightshade), which he drank and put in his eyes. As a result the active ingredient, atropine, was later purified and is now used to dilate the pupils for examining the eye. It is also used to treat stomach ulcers and to counteract the nerve agents used in chemical warfare.

Camphor is now best known as the smelly ingredient of moth repellents, but in Purkinje's day it was mixed with opium and given to children as a cough medicine. Purkinje's tests of camphor left him unconscious and entirely disorientated for days, while the reported effects on children were even worse. He concluded that children would be better off with the cough (and possibly moths) rather than the cure.

Thanks to Purkinje, physicians were alerted to the importance of fixing a safe dose for every drug and to the dangers of one drug amplifying the ill effects of another. When he wasn't putting his health on the line he studied lines, in the form of the ridges on fingertips, which eventually led to the forensic use of fingerprints.

In Victorian times arsenic and other potentially deadly poisons were readily available at every chemist's shop and poisoners were among the best known celebrities of the day. Francis Galton, who is now best remembered for founding the study of eugenics – selective breeding to improve the human race – spent his early days in charge of a pharmacy. To get a working knowledge of his wares he swallowed them. He began to work his way through them in alphabetical order but soon ran into trouble with aconitum and arsenic. He gave up after croton oil, one of the most powerful purgatives known to man.

Industrial workers were also exposed to a rich menu of toxic substances. There was an urgent need for an antidote. In 1813 a chemist called Bertrand swallowed arsenic mixed with charcoal and suffered no adverse symptoms. The charcoal had absorbed and effectively inactivated the poison. This result intrigued Pierre-Fleurus Touéry. Could charcoal neutralise other toxins? Nowadays charcoal is given to dogs embarrassed by aromatic wind, but Touéry spent two decades testing the consequences of giving dogs charcoal with various poisons. The medical establishment was doubtful about the value of his results so in front of a crowded gathering at the *Académie des Sciences* he swallowed ten times the fatal dose of strychnine – plus charcoal. The atmosphere in the lecture theatre was electric. Everyone in the audience expected him to collapse and writhe around on the floor, dying before their eyes. They knew of no antidote. Fortunately Touéry survived, but he had taken a tremendous risk because the absorptive properties of charcoal varied immensely: a poor batch would have done for him. Thanks to his courage we now routinely use charcoal to absorb noxious gases and to treat poisoning by alkaloids and similar drugs.

Seemingly unlikely compounds were found to be useful drugs. Anyone who has seen Henri-Georges Clouzot's wonderful thriller *Le Salaire de la Peur* (*The Wages of Fear*) will know not to shake a bottle of nitroglycerine, but medics were once convinced that it could soothe nervous disorders. In 1858 Doctor Field, an English physician, tasted nitroglycerine, an unstable high explosive produced from glycerine plus nitric and sulphuric acid. He immediately blanched and collapsed, convinced that his head would explode. His pulse faded to nothing, but his attendants brought him round with great difficulty.

Despite Field's close call a young physician called William Murrell later repeated the experiment. He was astonishingly casual, tasting nitroglycerine and then starting his normal routine of consultations. Soon, however, his head was throbbing and his heart pounding. It 'became so severe that each beat of the heart seemed to shake my entire body . . . the pen I was holding was violently jerked with every beat'. Nonetheless, he went on to dose himself and his patients with nitroglycerine numerous times more. The astute doctor noted that some of the effects were similar to those produced by a drug then used to dilate blood vessels narrowed by disease so he trialled it on his patients. Today nitroglycerine is the standard treatment to relieve the oppressive pains of angina.

Virtually all drugs were once taken by mouth. Then Enoch Hale Jr, while working at the Massachusetts General Hospital in Boston, had a seriously ill patient who couldn't swallow. But how else could a drug be delivered into his body? Hale realised that to inject a drug directly into the bloodstream was 'highly dangerous, or at best of very doubtful safety'. Many dangerous drugs are at least partially inactivated by the digestive acids in the stomach. Morphine, for example, is vastly more potent when injected than when taken orally. To bypass this 'safety valve' was risky but it was worth a try. So he injected castor oil, a mild purgative, into rabbits and when there were no obvious ill effects he injected some of it into his own vein. Soon he felt enfeebled with headaches and ever-worsening stomach cramps. His face muscles became partially paralysed and for hours he couldn't speak properly or eat. Even a month later he had not fully recovered. But the genie was out of the syringe and injection became an essential means of getting drugs as rapidly as possible to their site of action in the body.

A problem for drug testers is that a similar dose of the same drug may have very different effects on different people. A self-experimenter is only a sample of one. But if he survives and his experiments on animals and cell cultures go well, the drug may be deemed sufficiently promising to be trialled by volunteers. The point of these trials is first to ensure the drug is safe and then to establish that it is effective at combating a disease.

Many pharmaceutical firms once had in-house teams of volunteer testers but now they contract out to firms specialising in arranging drug trials. The drug trial business rakes in $24 billion a year. In Britain alone there are over 1,000 drug trials a year involving 100,000 or so volunteers. It has been estimated that at any given time around 50 million people worldwide are guinea pigs in clinical trials.

Volunteers are reimbursed for their time, often generously. But paying significant sums attracts those in developing countries who need the money. In the United States it is creating a growing band of 'professional guinea pigs'. It is estimated that there are around 10,000 volunteers for whom participating in clinical trials is their main income. One serial volunteer spent 500 nights in research facilities in five years. This is not good for his health. Flitting from one trial to another to maximise his income may also encourage him not to observe the thirty-day 'wash-out period' that ensures all traces of previous medication have gone before his body is challenged again. Flouting this caveat may result in a dangerous combination of drugs in his system. 'Career guinea pigs' are also less likely to own up to side effects that might get them dropped from the trial and reduce their fee.

Over a series of trials each new drug may be tested on 10,000 people before being licensed for clinical use. Most drugs never make it to that stage. Many fail the laboratory tests and even those that have showed early promise are often disappointingly ineffective when tested on people. Of the numerous anti-cancer drugs that pass trials on humans, only five per cent are ever approved for general use.

All drugs can have unpleasant side effects. Approval is based on balancing the risk of adverse effects against the benefits that the drug might bring. I have occasionally taken a non-steroidal anti-inflammatory drug. The accompanying leaflet listed no fewer than sixty side effects. Large-scale trials on humans enable the manufacturer to estimate how frequently particular ill effects may occur. For this drug one in ten patients might feel somewhat short-changed to get only a little dizziness, diarrhoea or a skin rash. One in a hundred recipients may be troubled by swollen ankles, stomach bleeding, difficulties in breathing and hepatitis. But Mr One-in-100,000 enjoys numerous ordeals including hearing loss, a swollen tongue, bloody urine, fits and night-mares, hair loss and impotence. All these were worse than the condition for which I was taking the tablets. There was, however, a reassuring footnote: 'Do not be alarmed by this list.' Too late: I was already a nervous wreck.

The worst problem occurs when a really serious effect only comes to light after the drug has been in use for some time. In the 1960s thalidomide, a sleeping pill, was found to have produced extreme birth defects if taken by pregnant women. It was such a potent substance that it could disrupt the development of any animal and even the embryos of plants were grossly malformed by it, but pregnant animals had never been tested. For several decades a synthetic oestrogen called diethylstilbesterol was prescribed to prevent miscarriages. In 1971 it was found to cause vaginal cancer in daughters of patients who had taken the drug, but in the United States it continued to be prescribed for some time as a 'morning after' pill. Recently a new painkiller caused heart

attacks and strokes, many of them fatal. In 2001 the manufacturer paid out $4.85 billion in compensation.

The purpose of initial drug trials on human volunteers is
to find unexpected side effects. These so-called 'phase one'
trials are the most risky and they usually involve only ten
to twenty people at a time. In March 2006 a German biotech
company hired Parexel, an American drug-testing firm in
London, to test their latest product. Eight volunteers were
chosen from respondents to an advertisement on the
internet. They were contracted to stay at the clinic for two
days during the trial and then attend several times later on
for check-ups. They would be paid £2,000 each.

The healthy young volunteers were infused with the drug
one after another. They had been warned that they might
experience slight headache or perhaps some nausea. But as
the last man was being injected one of the others began to
feel ill. Soon he was screaming with pain. He tore off his
shirt, shouting that he was burning; then he convulsed and
collapsed. One by one the others followed suit, writhing on
the floor and begging the doctors for help. Within minutes
the calm of the clinic had given way to the chaos of a
Victorian asylum. Two of the volunteers watched in horror,
waiting their turn to collapse, but luckily they had been
given a placebo and were safe. The other six were far from
safe. Within ninety minutes they were having difficulty
breathing and their pulse rates had doubled. This was an
unprecedented emergency. Now comatose, they were

rushed to hospital. The staff at the intensive-care unit had never seen anything like this before. Nine hours after taking the drug the men were unable to breathe unaided and were facing multiple organ failure.

It was twelve hours before relatives were allowed to see them. They couldn't believe their eyes. The face and neck of twenty-one-year-old Ryan were swollen to twice their normal size; his body had inflated to that of a 190-kilogram (thirty-stone) man and his skin was purple. The partner of twenty-eight-year-old Nino said he looked like the Elephant Man. The doctor confided that Nino and Ryan 'needed a miracle'.

Ryan, the most badly affected, was in a coma for three weeks, suffering with heart, kidney and liver failure as well as septicaemia and pneumonia. His fingers and toes were black with dry gangrene and as hard as stone. Several were amputated. He was not released from hospital for three and a half months.

So what went wrong? The drug being tested was TGN1412, a monoclonal antibody developed by a company called TeGenero (which translates creepily into 'I create you'). Monoclonal antibodies are artificial versions of the body's natural antibodies that fight off disease. The Nobel Prize was awarded to the British scientists who developed them and they are the great hope for twenty-first-century medicine. Dozens of them are now in medical use and many more are being tested. Herceptin, for example, treats breast cancer by attaching to malignant cells and stopping them from multiplying, thus curbing the tumour's spread.

TGN1412, on the other hand, could also *stimulate* cells to *divide*. It targets white blood cells called T-cells, a major player in the body's defences. The hope was that TGN1412 would boost the body's immune system into action by creating reinforcements to attack the rapidly proliferating cells of lymphatic leukaemia.

It would seem sensible to test a new drug on fit and healthy volunteers who could cope with any adverse side effects, but in this case it was a disaster. Cancer victims are under siege and have weakened immune systems that are desperately in need of a boost, but healthy people do not. Inside the volunteers' bodies TGN1412 stimulated a chain reaction, releasing millions of cytokines, the immune system's messengers, which in turn triggered T-cells to liberate even more cytokines. This caused a 'cytokine storm', a deadly overreaction in which the immune system attacks the body, inflaming blood vessels and consuming internal organs. No wonder the victims thought they were on fire.

When humans are first exposed to an untested drug it is usual to dose a single volunteer to begin with, then observe how he gets on before exposing anyone else. The second volunteer might be tested several hours, days or even weeks later. Those testing TGN1412 were dosed within only a few minutes of each other, thus putting all six of them in danger.

TeGenero claimed that there was nothing in the laboratory experiments to suggest there might be a violent reaction to the drug. But what occurs in cell cultures exposed to a drug is not a reliable indicator of what might happen inside the human body. Nor were experiments on animals a good guide in this case. The TGN1412 antibody was made up mostly from *human* protein and there are biochemical reasons to suppose that the response might therefore be far greater in humans than in rats or rabbits.

The after-effects were not entirely a surprise. Parexel had indicated in the contract signed by all the volunteers that cytokine release was a possible consequence of taking the drug, and TeGenero had notified the Medicines and Healthcare Products Regulation Agency, who authorised the trials, that there was a risk of cytokine release and that a similar drug had induced a bad reaction.

TeGenero went into bankruptcy. A government inquiry into what had gone wrong made many recommendations to prevent another such incident but failed to apportion blame. There had never been such a disastrous trial before, nor has there been one since.

The victims still suffer from the physical and psychological aftermaths of their traumas. One of the participants, David, has memory loss and the early indications of an aggressive cancer. Nino has numerous tender lumps emerging all over his body. His partner, Myfanwy, naturally wishes that he had never involved himself in the trial. At the time she had voiced her qualms, but he had reassured her – it was not just the £2,000 fee, he was testing a cure for leukaemia and helping mankind.

What would we do without such willing volunteers? Myfanwy thought that we should be thankful to these men 'for having the courage to help bring medicines to us all'. Indeed we should.

Mandrake root whose anatomy suggested it would work wonders in love potions.

Lovely Grubs

'The shrivelled objects, resembling pieces of a horse's hoof, were soaked all day and boiled all night, by which time they looked like large black garden slugs' — Description of sea cucumber served to dinner guests by Frank Buckland

The most complex substances that we swallow are not drugs but the organic compounds we affectionately call food. In olden times we were far less finicky than today. Anything that swooped, scampered or slithered was considered to be grub. Until the nineteenth century English recipes included such ingredients as seal and squirrel, and the wealthy consumed swans or roasted dolphin in a porpoise sauce. No crane, lark or song thrush was safe from the pot. There is rather little meat on most British birds, so no wonder it took four and twenty blackbirds to fill a single pie.

The Victorians changed all this. A famous zoologist at

Edinburgh University taught that civilised folk should confine their comestibles to cultivated or husbanded stock created for the 'special use' of humanity. Refinement demanded dietary discipline. It was only the savage who consumed indiscriminately. The evidence was overwhelming; when the heathen Hottentot was baptised by missionaries he became nauseous at the sight of zebra meat, which he had relished all his life. Even our French neighbours craved frog's legs, snails and horse-flesh. Orientals, no matter how ancient their civilisation, were treated with suspicion and 'animals that in England were looked on with disgust . . . are by the Chinese regarded as delicacies'. Who else would eat grubs, earthworms, rats and, worst of all in British eyes, dogs and cats?

But there was a desire for the exotic. Explorers had been roaming the globe gathering useful animals and plants that might be exploited back home or in the colonies. The seeds from a single African coffee plant were transported to Brazil by the Portuguese and bananas from Asia were being cultivated in the West Indies. Captain Bligh, of *Bounty* fame, also introduced breadfruit into Jamaica from Tahiti. Some food plants would even grow in Europe: wheat had been imported from Asia, barley from the Middle East, maize, potatoes and tomatoes from the Americas, and 'French' beans from Canada. Most of the great botanic gardens were set up as 'gardens of acclimatisation' to cultivate introduced plants and perhaps even 'persuade' some semi-tropical species to cope with the European climate.

Many farm animals had also been introduced: cows, chickens and turkeys had been naturalised long ago. Roving naturalists such as Joseph Banks, who accompanied Captain Cook on the cruise of the *Endeavour*, tasted many of the exotic animals he discovered. He was the first white man to taste kangaroo and boasted, 'I have eaten my way into the animal kingdom farther than any other man.' But that was before Frank Buckland came along.

Frank was the son of William Buckland who was Oxford's first Professor of Geology and became Dean of Winchester. From childhood Frank was fascinated by animals and his collection of wildlife grew to include guinea pigs, doves, hedgehogs, dormice, frogs, tortoises, marmots, snakes (including a venomous adder), monkeys, chameleons, a jackal and lots more besides. They all seemed to be escapologists. His eagle soared around inside the chapel during a service, the cat lodged in one of the organ pipes transforming well-loved hymns into deathly drones. His baby bear raided the local sweet shop, 'putting the village in uproar', before charging into the chapel and dumbfounding the reader of the day's lesson. Florence Nightingale suggested that the unruly bear should be hypnotised. This desire to keep animals in a state of intermittent incarceration remained throughout Frank's life. He eventually had a 'studio' full of wild livestock and woe betide anyone wearing a flouncy skirt or a coat with flying tails who passed close to the drunken monkey's cage. His huge Turkish wolfhound escaped and spied a lady and her little dog in a neighbouring house. The wolfhound leapt through the open window and within moments the lapdog had lapsed for ever.

Thanks to his father, Frank's love of strange animals extended to eating them. Dean Buckland had boasted that when present at the disinterment of Louis XIV he sliced off part of the embalmed heart and had it for tea. Buckland's dining table groaned beneath dishes of horse's tongue, ostrich, frogs, snails and rats. A house guest later groaned that he had not enjoyed the crocodile for breakfast.

It's not surprising that in later life Frank's cook would be asked to serve such indelicacies as hedgehog and puppy. John Ruskin wrote to say he was sorry to have missed the toasted mice. Meals rarely passed without interruption from a flying fox or a wild hare. On one occasion a boot repeatedly flew across the floor as if it were a prop in an express séance. A meerkat had mistaken it for a burrow that it was trying to excavate. Another time a distinguished cleric was propelled rapidly backwards from the table by a river hog wedged between his chair legs.

Frank was co-founder and secretary of the Acclimatisation Society whose aim was to encourage the use of imported animals so that the British public should not be deprived of the joys of emu stew or fried wombat. Frank declared that 'I do not mean to let slip any opportunity of increasing the supply of food for the people.'

It was the members' task to table-test novel foodstuffs. At the first official dinner of the society in 1862 not all of the numerous courses received universal acclaim. They began with three Chinese soups:

Bird's nest soup from the regurgitated seaweed slime that holds the nest together. Verdict – 'Gelatinous with a very peculiar flavour.'

Sea slug soup – 'Tasting something between a bit of calf's head and the contents of the glue pot.'

Deer-sinews soup – 'After a monstrous deal of boiling, good eating, but glue-like.' Frank decided that: 'When I next entertain a Chinaman, I'll give him . . . sixpennyworth of carpenter's glue.'

Semoule soup from Algerian wheat – 'More fitted for invalids than for ordinary table. Reminded one of the porridge the giant was eating when Jack the giant-slayer killed him.'

Some entrées fared no better; the kangaroo ham was 'dry and too highly salted' and kangaroo steamer had 'gone off'.

Four years later 160 diners attended a feast to popularise horseflesh. Every course from the soup to the dessert jelly was of equine origin. The reviews were not encouraging: 'Simply horrible.' 'Resembled the aroma of a horse in perspiration.' Frank was ill all the next day and concluded that: 'In my humble opinion hippophagy has not the slightest chance of success in this country.'

Funny food caught on and as an undergraduate at Cambridge Charles Darwin joined the 'Glutton Club' that convened weekly for esoteric eating. According to Darwin the cod's tongue was fine but its liver was 'not good' and the brown owl was 'indescribable'. Almost all the 'gluttons', except Darwin, rose to high rank in the church.

When Frank Buckland heard that a panther had died at the Zoological Gardens he asked for some samples of the beast. So they dug it up from its grave and dispatched parts for examination. Frank examined them with his taste buds, but the chops 'were not very good'. The director of the zoo, thinking that Frank's interest was anatomical rather than gastronomic, asked if he would determine the cause of death of future fatalities. Frank could not resist such a mouth-watering opportunity.

He was competent to carry out post-mortems, having qualified as a surgeon, and had been house surgeon at St George's, the hospital where John Hunter had accepted the same post eighty-four years earlier. It had changed surprisingly little. Surgeons still operated in coats encrusted with coagulated blood – a sanguinary souvenir of past patients – and the wards were redolent with the aroma of gangrene. The nurses were largely untrained and illiterate. When Frank asked one to read the label on a jar, she ventured: 'Two spoonfuls to be taken four times daily.' It actually said *To be applied externally only*. Frank's cases were mostly failed suicides and 'scaffold incidents' where the gallows had given way under the gyrations of the choking felon. After a typically 'delightful day of dissection' he snacked on chicken's brains and read a 'lively dissertation on inflammation of the bowel'. Such was his enthusiasm for dissection that it was said that 'elderly maidens called in their cats as he passed'.

The cadavers from the Zoological Gardens were far more varied. A young visitor who called while Frank was conducting an animal autopsy described a large cadaver on the table and Frank frequently stopping to slurp from a bowl of stew beside the corpse. 'Have some?' he asked invitingly.

His staff became accustomed to opening unusual parcels but had several frights, as when three live badgers popped out, or a scorpion arrived in a jeweller's gift box. The animals' remains invariably ended up in the oven. Indeed, sometimes they were cooked before Frank had examined them. 'Directly I am out of the way,' he protested, 'if they *look* good to eat, they are cooked; if they stink, they are buried. What am I to do?' The upside was that he got to eat bison, giraffe, viper, rhinoceros pie, boiled elephant's trunk and whole roast ostrich.

Buckland would be delighted that his culinary habits live on in 'Buckland Dining Clubs' in Birmingham and Helsinki, and that antelopes and okapi now roam England's country parks just as he predicted, but he would be disappointed that they are not being raised for food. Yet for all his enthusiasm for animal importations his most important transplant was an export. He was appointed the Government Inspector of Salmon Fisheries and did a wonderful job. Although he had no scientific training and was almost innumerate, he was genial and straight-talking so fishermen, water bailiffs and even poachers warmed to him and willingly gave information. He reported to government on overfishing, water pollution and fish diseases. Another problem was the demand for water power. Where there was a water mill there was a dam. The River Severn had seventy-three weirs and each one was a barrier to salmon migrating up river to spawn. He supervised the introduction of fish 'ladders' to enable the fish to by-pass the dams.

Frank also drew attention to the lack of knowledge of the breeding and biology of both freshwater and sea fishes, and stressed the need for research on exploited species. He was ahead of his time and was one of the first to propose that this work should be the government's task and not one for individual researchers. Better understanding and improving water quality would greatly increase the catch, but Frank realised that the stock could also be augmented by artificially cultivating the fish. He expelled eggs from a pregnant female by gently squeezing her sides and mixed it with milt from a male. In this way he hatched 30,000 salmon and trout eggs in his kitchen sink. He provided a thousand trout eggs which were exported to the Antipodes where they gave rise to all the brown trout now inhabiting the waterways of Tasmania and New Zealand.

It may come as a surprise that he ate something as palatable as his experimental fish. Tasting your research is not uncommon. I knew a marine biologist who studied plankton and in the evening offered me plankton sandwiches. Thomas Hunt Morgan was awarded the Nobel Prize for confirming the rules of genetic inheritance and mapping the location of genes on their chromosome. His experiments involved cross-breeding tiny fruit flies. To get to know them better he ate their maggots. They tasted like breakfast cereal. There are stranger things to eat for science. Lazzaro Spallanzani investigated the process of digestion by swallowing food in linen sacks and later regurgitating them for examination. Surely the ultimate high-fibre diet.

Except for his diet, Frank Buckland was no experimentalist but was practical and full of ideas. Unfortunately, for every good idea he had a daft one. To fatten fish more rapidly he suggested dangling a horse's leg or a bunch of dead rats from a tree branch overhanging a fish pond. Over time the decaying meat would fall into the eager mouths of the fish below. He advised a woman distraught at the death of her pony that its hooves would 'make good inkstands' and its preserved ears 'nice holders for spills'. He was against children wearing shoes for walking because it makes leather soles thinner whereas going barefoot makes the soles of the feet thicker. He had admired the feet of Scottish 'fishergirls' whose skin was 'as thick and hard as the foot of an elephant'.

Frank Buckland was the most entertaining writer of popular-science articles of his generation. For decades he turned out tales of natural and unnatural history, such as 'Elephant bones from the bottom of the North Sea' and 'Buoys made from inflated dogs'. He was intrigued by giants, perhaps because he was only 4 feet 6 inches (137 centimetres) tall. He described a fish called *Chimaera monstrosa* in typical style: 'Couch describes its habits as nocturnal; he is doubtless right, for such a hideous fish can hardly dare to show itself in the day . . . coming round the corner on a dark night, [it] would be enough to frighten any ordinary fish into a fit of hysterics.' This species *is* weird-looking. It resembles a dark, elongated carrot with ears, and inspired a children's rhyme:

> My name is *Chimaera monstrosa*,
> My body gets grosser and grosser,
> From the tip of my tail,
> which is merely a flail,
> To my head with the face of a grocer

To peckish readers Frank offered recipes such as slug soup: 'The great grey and black slugs, I believe, make the best soup . . . if boiled till a rich firm jelly.' He also praised the culinary potential of the capybara, blind to the possibility that the British public might not be tempted by a rodent resembling a rat on steroids.

Frank had an encyclopaedic palate. While visiting a cathedral to investigate a manifestation of 'fresh martyr's blood', he did indeed find wet patches on the floor. He tasted one and announced, 'Bat's urine.' I wonder how many samples he had tasted to distinguish bat's urine from, say, a rat's or a bishop's?

Even he found some creatures inedible. Among his rejects were stewed mole, bluebottles and earwigs, which were 'horribly bitter'. Disappointingly, the head of porpoise tasted like the 'broiled wick of an oil lamp'. Perhaps he would have looked more favourably on them had he been *in extremis*, stranded in a place where only earwigs and lamp wicks grew. Such a possibility (more or less) was considered by Francis Galton.

Galton, a half-cousin of Charles Darwin, dabbled in self-experimentation, even trying to suppress automatic bodily functions. On one occasion he succeeded so well that he almost suffocated.

Galton trained to be a doctor, but his education was interrupted when he took a 'gap year' to travel abroad. No sooner had he returned to his studies than his father died, leaving him sufficiently wealthy to travel to more distant places. A two-year-long expedition to the unexplored regions of south-western Africa involved danger and deprivation, but Galton believed that 'alternate privation and luxury is congenial to most minds'. He encountered hostile native tribes but pacified them by the sheer force of character. For

his anthropological studies he developed a method of measuring busts and hips from a distance using a sextant, although
most men can do this without any instrumentation. His
reports of the trip earned him the Gold Medal of the Royal
Geographical Society and led to his election as a Fellow of
the Royal Society while still in his early thirties.

In 1872 he published an explorer's handbook called *The
Art of Travel*. It contains sections entitled 'Wholesome Food,
procurable in the Bush' and 'Revolting Food that may save
the Lives of Starving Men'. The latter is full of helpful tips,
e.g. if the water is suspected of being poisoned get a cat or
dog to taste it first. I can see that the explorer might be
accompanied by native bearers, but a cat? One learns that
'Carrion is not noxious to a Starving Man'. Apparently, a
diet that would cause a healthy, well-fed chap to fall dangerously ill is absolutely fine for one who is starving. In this
condition 'carrion and garbage of every kind can be eaten
without the stomach rejecting it'. Rotting carcasses are also
easy to find: just follow a jackal, your friendly guide, or look
for circling crows and vultures sitting on trees. It's perhaps
wise to ensure that the vultures are gorged prior to your
arrival.

Birds should be skinned because their 'rankness lies in
their skin'. Yet 'all skins of any kind are fit and good for food;
they improve soup ... or they may be toasted and
hammered ... Many a hungry person has cooked and eaten
his sandals.'

A problem arises if you manage to capture an animal. It makes a great feast for today, but what about tomorrow's dinner? A useful find would be an animal with deciduous parts. Galton noted that bush ticks eroded the base of oxen's tails to the extent that the tail dropped off, reminding him that oxtail soup is 'proverbially nutritious' – with the guilty ticks as a side dish, perhaps. He appreciated that you might be too weak to carry a carcass so he suggests that you use the old Abyssinian trick of keeping the animal alive and taking it with you, just slicing off what you want each day. He doesn't say how to keep the poor animal's spirits up during the trek.

Galton found that locusts and grasshoppers 'are not at all bad'. The great virtue of insects as food is that they are never in short supply. When the famous biologist Jack Haldane (whose extraordinary exploits will be narrated later) was asked what his studies had told him about God, he replied 'that He had an inordinate fondness for beetles'. Certainly the vast majority of animals that the starving explorer encounters would be insects and it would be foolish to disdain them. Some bugs contain seventy per cent protein, more per gram than meat and less fatty, as well as being rich in vitamins and minerals. Although nutritious, insects are not invitingly pack-aged, a serious failure in the Deity's marketing department. They are, however, consumed throughout the developing world. Locusts are eaten raw, roasted, fried, jellied and mashed, and are a seductive combination of a crisp exterior and a creamy filling. But remember to remove the legs first as they get stuck between the teeth. If you're wanting more than just a snack, then cockroaches fit the bill. For the calorie counter, ants and termites also have a worldwide following, but the tyro collector should remember they deliver a hefty bite for their size and inject formic acid as an irritant. For this reason they are best eaten cooked rather than live or raw. The best ants' eggs are known invitingly as 'sour ant' in Thailand. When bitten they pop out a cream tasting somewhere between camembert and hard-to-bear.

A few westerners are also insectivores. There is even a 'Food Insects Newsletter' produced by the University of Wisconsin. Frank Buckland would have been in his element at the centennial dinner of the New York Entomological Society in 1992. The courses included:

Spiced Crickets and Assorted Worms
Waxworm and Avocado Roll
Worm Fritters with Plum Dipping Sauce
Cricket and Mealworm Sugar Cookies

The queasy should be reassured that the worms were, of course, not really worms at all. They were grubs. Although I *have* seen a recipe for worm patties, which requires almost a kilo of ground earthworms.

Unbelievably, we all eat about a kilo of insects every year, mostly because these pesky critters can't be kept out of food processing. The US Food and Drugs Administration allows up to 450 insect fragments in every kilo of wheat flour, 225 insect fragments or four and a half rodent hairs for every 225g of macaroni, sixty insect fragments or one rodent hair per 100g of chocolate, one maggot or five fly eggs in 250ml of citrus juice and one pellet of rat excreta per sub-sample of popcorn. In most pizzas, bangers and crisps there is an additive called cysteine. It comes from human hair.

The problem for the traveller is how to know which creatures are edible and which are poisonous. Some animals are venomous and best not approached or they might get you instead of the other way round. Frank Buckland almost died from a snake bite, but the snake didn't bite *him*. He was poisoned while dissecting the snake's dead prey. Fortunately he didn't eat it. I wonder how many of our ancestors died testing whether a succulent-looking spider or shiny berry was safe to eat. The risk is encapsulated in Jonathan Swift's comment that: 'He was a bold man that first ate an oyster.'

Everyone knows that some toadstools are lethal, but which ones? Perhaps their names hold a clue: Devil's Boletus, Destroying Angel, Death Cap. Seemingly innocuous plants or their products can also be dangerous. Nutmeg is a poisonous narcotic. Even a small amount of chocolate is toxic to many animals, although it takes eleven kilos to kill a chocaholic human. Many plants contain deadly poisons such as cyanide, strychnine and prussic acid. Tomato and potato plants come from the same family as deadly night-shade and are just as poisonous. The only safe parts are the ones we use: the fruit of the tomato and the tuber of the potato. Noxious chemicals in plants deter grazers, but some animals have developed immunity to the toxins contained in their main food source. This sometimes has disastrous consequences for humans.

When the Americans recaptured the island of Guam from the Japanese in 1944 a navy doctor discovered that the main cause of death among locals was a devastating brain disease resulting in paralysis, dementia and death. Outsiders didn't get it — so what was the cause? Native islanders made flour from the nuts of the cycad plant, which contain a potent nerve poison. Surely that was the culprit? But, puzzlingly, the traditional way of preparing the flour removed almost all of the poison. However, they also ate local bats called flying foxes. The bats fed on the cycads and their bodies accumulated the nerve agent so that over time their tissue contained four hundred times more poison than would be found in a tonne of processed cycad flour. It is common-place for both animals and plants to accumulate dangerous substances within their tissues to very high levels without showing any adverse symptoms. Thus a seemingly healthy animal may be deadly if eaten.

Neither Frank Buckland nor Francis Galton fully appre-ciated the risks they were taking. Frank knew that some animals might be poisonous, but his test was to taste them.

The garfish has sinister-looking green bones and was thought to be toxic. To ensure that it wasn't Frank ate a dozen of them for supper. Others deliberately eat animals that are potentially deadly.

While exploring various Pacific islands Captain Cook, against the advice of the naturalist on board, dined on puffer fish and became very ill. Fortunately he had consumed very little and it was one of the less toxic species. In Japan the most poisonous one is an expensive delicacy.

The *fugu* fish is not merely food: it is a drug, a ritual and a dining event. Raw *fugu* flesh is at the zenith of epicurean eating. The platter is an arrangement of a hundred or more transparent slivers of pale flesh arranged like the petals of a chrysanthemum, or the outspread wings of a crane. The effect is breathtaking. How can anyone bear to deface it by taking the first piece?

Its preparation is only permitted by state-licensed chefs who spend four years training to ensure that their customers may dine without dying. The most toxic organs – the liver, ovaries, intestine and skin – must be removed completely and the remaining flesh washed thoroughly to remove all traces of the toxin.

The *fugu*'s poison is a neurotoxin that stops the nerves from transmitting. It is one of the most deadly compounds on Earth, over twenty-five times more powerful than the arrow poison curare, and 10,000 times more lethal than cyanide. The amount needed to kill a person would fit on a pinhead. It is not a pleasant death. A tingling sensation gives way to burning pain and stomach cramps followed by muscular paralysis and progressively greater difficulty in breathing. There is no antidote and, as with curare, the patient is aware of everything that is happening but is unable to move or speak. The lucky ones die within eight hours.

Even though every effort was made to remove the toxin, gourmet diners sometimes begged the chef to leave just the tiniest trace to provide that tingling euphoria they craved. The result was sometimes that recounted in a traditional verse:

> Last night he and I ate *fugu*,
> Today I help carry his coffin.

In 1975 Japan's leading stage actor, who was officially designated a 'living national treasure', died from *fugu* poisoning. Subsequently a ban was imposed on the preparation of *fugu* liver. At that time the death rate was running at over twenty people a year.

But the Japanese gourmet seeks more than mere safety, he wants the thrill and danger that a trace of poison provides, for a *fugu* without its poison is said to be like a samurai without his sword. An old verse states the dilemma:

> Those who eat *fugu* are stupid,
> But those who don't are also stupid

NB: if you are a daring diner and wish to order *fugu* in a Japanese restaurant, make sure you don't ask for *fugo*, which is a bomb suspended from a balloon and should not be swallowed in any circumstances.

Frank Buckland medicating a poorly porpoise.

A Diet of Worms

'The loathsome parasite is both an exquisite example of adaptation to environment and ethically revolting.' — The Bishop of Birmingham

Far less beautiful things than flakes of *fugu* lurk within us. It is said that if all the tissues of a human body should magically vanish, the form of the person would still be clearly discernible from the distribution of their parasites. We are a walking menagerie of horrors. Parasites can infest every organ of our body from the lungs and heart to the eye and brain. None of our secret places is safe from habitation — there are even tiny worms that inhabit our eyelashes.

Most of our internal lodgers do little harm — after all, it is not in a tenant's best interest to demolish the boarding house. But that doesn't make them desirable. There are differences of opinion even among biologists. When Marlene Zuk, an American zoologist, was asked to name her favourite parasite, she chose a worm that takes over the brain and body of an unlucky grasshopper and becomes so large that it bursts open the insect to emerge. Clearly she had not sought the grasshopper's opinion. The curator of parasitic worms at the British Museum plumped for his giant pickled tapeworm. On the other hand, Jack Haldane wrote, 'Did God who made the lamb make thee? The same question applies to the tapeworm and . . . would clearly postulate a creator whose sense of values would not commend him to the admiration of society.'

For most people the tapeworm occupies the more repellent end of the spectrum. It's a long, flat alien found in the human gut where it can become a long-term resident. One man harboured the same tapeworm for thirty-five years, and the biggest specimen ever to be 'passed' by a human had clearly been inside him for a while as it was thirty-nine metres long.

It has a pin head with suckers and thornlike hooks to anchor itself in the wall of the host's intestine. The head manufactures a succession of broad, flat segments until there may be as many as a thousand. Production never stops and five to ten segments are shed from the 'tail end' every day. Each contains 100,000 eggs. The host has inadvertently become the egg depositor for a battery-reared flatworm that he feeds on a continuous slurry of pre-digested food.

In 1855 Dr Küchenmeister tried to resolve the question of whether humans acquired their flatworms through poor hygiene and ingesting the eggs that they (or others) had deposited, or from pigs that had swallowed the eggs, which encyst as 'bladderworms' in the hog's muscles. There could be as many as 3,000 bladderworms in half a kilo of pork.

If Küchenmeister was to attract volunteers to test the possibilities, he thought it might be wiser not to mention that bladderworms, once swallowed, can lodge in the liver, muscles and eye (causing blindness), or in the brain – resulting in epilepsy or insanity. Fortunately he was acquainted with the Duke of Saxe-Coburg who offered doomed convicts for his experiments. Küchenmeister served one condemned man bladderworms in blood sausage and soup. The prisoner complained that the soup was cold, not appreciating that hot soup might kill the worms and ruin the experiment. After his execution a post-mortem revealed ten tiny tapeworms in his gut. Surprisingly, this inspired a volunteer to swig lukewarm milk laced with bladderworms. Three months later, to Küchenmeister's delight, the volunteer began to pass tape-worm segments. This test was repeated with a similar result on several more volunteers, but the final proof came from another convict whom Küchenmeister infected via worm-spiked sausage spread on a bread roll. When the convict was executed four months later, his gut was found to contain several full-grown tapeworms. Clearly humans could be infected from undercooked contaminated meat.

In many countries tapeworms have been deliberately

disadvantaged by the population not eating sacred cows or unclean pigs. With better hygiene and food inspection tapeworms have virtually died out in western countries, but our liking for very rare steak keeps a few in business.

It was common practice to carry out dangerous experiments on convicts even into the twentieth century. They were considered the dross of humanity, who were generously being given an opportunity to perform a small service for society. There was, of course, no requirement to ask their permission. What is surprising is that Küchenmeister also persuaded willing volunteers to take part, for at that time there was no reliably effective treatment to rid the body of tapeworms.

Having hungry internal residents stealing your food after you'd eaten it was believed to make you thinner. Victorians marketed 'tapeworm tablets' for figure-conscious ladies. The logic behind this was that the poor were thin and most of them had parasites. When I was a child, any lad with a hearty appetite attracted the 'amusing' comment, 'You must have worms.' Parasite-induced anorexia is now a well-known phenomenon, but it's not attributed to the theft of digested food. Appetite can be suppressed by natural chemicals such as cytokines acting on the brain. Parasitic infection stimulates several responses in the host, including cytokine production, which induces a decline in appetite and consequent loss of weight.

Tapeworms are relatively benign as internal parasites go,

but some people are hypersensitive to their presence and suffer nervous disturbance or may even die. The biggest tapeworm to infest humans comes from eating raw fish and the Japanese take a weekly tablet to purge their innards of recent recruits.

A variety of parasites jostle to occupy the human gut. The most common is the large roundworm, which is estimated to infect over 1,000 million people. It resembles a pallid earthworm and can reach forty centimetres in length. In contrast to the relatively languid tapeworm, the agile roundworm is forever stretching, coiling and twisting. It may be present in vast numbers: 5,000 have been found in a single person. They form writhing tangles that can completely block the bowel or the ducts from the gall bladder and pancreas, causing intense pain.

Each female roundworm sheds 200,000 eggs which exit with the host's motions. If personal hygiene is not meticulous, a few eggs may be swallowed to sustain the infestation. These eggs hatch in the intestine and the small juveniles embark on an extraordinary odyssey around the host's body. They bore through the gut wall and then into a blood vessel to be carried to various organs in the body, including the heart, where they lodge to cause problems later. Those reaching the lungs burrow through into the air passages to be coughed up into the mouth and swallowed again. On entering the intestine for the second time they stay to grow and flourish. This grand tour of the body damages all the membranes that the creatures pass through and in pigs they frequently cause the deaths of large numbers of young animals from porcine pneumonia.

It's probable that the occasional coughing up of a round-worm caused momentary social embarrassment and also gave rise to legendary accounts of snakes living in the human stomach. There are vivid descriptions of vipers being vomited up, sometimes augmented by a reflux of lizards, frogs, toads and even mice, cats and dogs.

The biology of roundworms was worked out without resorting to self-experiment, but not without risk. Parasitologists take great care to avoid becoming infected, although a colleague of mine once shared a laboratory with someone studying large roundworms. She washed her hands scrupulously after handling the worms, but as she was in a wheelchair she first had to wheel herself over to the sink and then wheel herself back to the bench. Many researchers reacted to some of the substances exuded by the worms and suffered severe allergic reactions such as dermatitis and asthma. Despite this, most of them persevered with their research but some were completely prostrated by any contact with the worms.

Blood flukes are flatworms and are much smaller than the large roundworm but more dangerous. The male has a sucker to attach to the host's flesh. The female lives perma-nently embraced in a fold along the length of the male's body.

In 1908 Dr Claude Barlow became interested in flukes. It is to be expected that missionaries have a desire to convert people but Barlow, while serving as the medical officer at a church mission in China, converted himself. Having witnessed the severe distress and debility that the widespread infection by parasitic flukes inflicted upon peasant farmers, he trained to become a parasitologist.

Returning to China, Barlow realised that the locals were probably reinfecting themselves by using their faeces as fertiliser on the land. The larval flukes were so tiny that they could easily be ingested without the recipient knowing. No wonder fluke infection was endemic throughout rural China. He wondered whether the adult flukes might also be a source of infection, so he swallowed some taken from an infected patient. Unsurprisingly, it was 'a nauseating experience', even though he did it in the dark so he couldn't see what he was putting in his mouth. Barlow then observed what emerged from his other end, but found nothing remotely parasitical. He assumed that the flukes had been destroyed by his digestive enzymes but failing to see this was a lucky break, the dogged doctor persevered. He took an antacid followed by flukes for starters plus a main course, and he was overjoyed to see that he was passing fluke eggs. He continued to do so for a year before getting rid of the parasites with drugs.

Fourteen years later Barlow feasted on flatworms once

again. At the time he was working in Cairo. A pathologist had claimed that Egypt would never be an important country until the flukes draining the people's energy were controlled. He was referring to blood flukes that give rise to schistosomiasis or 'big belly'. It was the second-most common disease in the world and it still affects 200 million people. The parasites can survive in the human body for ten or even thirty years, but are often fatal to their hosts long before that.

Like many parasites the blood fluke has a complicated life history. It spends part of its existence inside freshwater snails. Paddling in contaminated paddy fields or washing in a stream is sufficient to become infected. In 1944 there was concern that Allied troops fighting abroad and having contracted schistosomiasis would bring it back home when they were demobbed. A senior medic declared that: 'The establishment of schistosomiasis in North America is a serious possibility.' Barlow wondered whether the snails back home would be susceptible to infection by blood flukes and would therefore provide the essential second host. He tried to import American snails into Egypt for his experiments, but most of them died in transit. If the snails couldn't be brought to the flukes, the parasites would have to be taken to the snails. The container in which they were shipped was Barlow himself. Although more expensive than postage, this method would not require an import licence.

Experts rightly dismissed his plan out of hand because potentially it could endanger his health. Blood flukes cause severe dysentery, anaemia and emaciation. They produce so many eggs that thousands or even millions of flukes remain in the host's body, causing widespread inflammation that stops the flow of blood to vital organs such as the bladder, liver, lungs and heart. They are lethal.

This didn't discourage Barlow. Over three weeks he dosed himself four times and also infected a baboon called Billy (after Herr Bilharz who had first described schistosomiasis). Three weeks later Barlow and Billy took a plane to the United States. During the flight Billy escaped and terrified the passengers.

Barlow was already having sweats and dizzy spells, and it would get much worse. It was known that a parasite's eggs emerged in the host's urine or faeces, but not through the skin. To find the parents of the eggs, Barlow had a piece of skin the size of a luggage label cut from his body. The incision went deep but he stoically refused a local anaesthetic in case it disturbed the parasites. The biopsy revealed adult blood flukes in the skin. New symptoms appeared frequently and each was worse than the last. Barlow should never have swallowed these flukes, they were far more aggressive than those he had ingested in China. He must have wondered what it was that first endeared him to parasitology.

Barlow was now suffering severe night sweats and

began to pass blood. He had almost more pain than he could endure. He got little sleep because he had to urinate every twenty minutes. His condition got progressively worse. After being confined to bed for three weeks, his temperature soared to 40° C. He was relieved to still be alive.

His morale was not helped when Billy the baboon, who had become a dear companion, died of schistosomiasis. Yet he continued his ordeal even though he was passing 12,000 eggs per day and had blood in his urine the whole time.

Eventually Barlow agreed to start a course of drugs to clear the infection. He returned to work in Cairo but got no better, so he sought treatment at an Egyptian hospital specialising in parasitic infections. He was injected with antimony, which is toxic. The flukes that caused schisto-somiasis are notoriously difficult to remove from the body so a powerful remedy had to be used. Antimony was, and I believe still is, the most effective agent. But it can result in rashes, diarrhoea, pains in every muscle and joint, and more rarely sudden death from heart failure or total collapse of the circulatory system. It damaged Barlow's heart. He began to vomit and felt nauseous all the time. But it worked. Eighteen long, miserable months after he had deliberately infected himself, he stopped passing eggs.

All this suffering left Claude Barlow too weak to carry out the experiments on snails that had been the purpose of his ordeal. Instead, a colleague tried to infect American snails with the flukes, but failed.

In the late nineteenth century there was a Europe-wide epidemic of miners and construction workers dying from 'miner's anaemia'. In 1881 an army of labourers dispersed after constructing the nine-mile-long St Gotthard tunnel through the Alps. They spread the disease from the tin mines of Cornwall in the north to the sulphur mines of Sicily in the south. All the victims were found to be infested with hookworms.

The intestinal hookworm is tiny, but it has vicious, hooked 'teeth' that bite into the gut wall so that it can suck blood from its host. One or two worms would do little damage but some miners contained thousands of them. With so many bloodthirsty inmates, no wonder the men became anaemic and died. An infected person passed uncountable numbers of eggs and it was assumed that others became infected by inadvertently swallowing some. But improved hygiene in mines failed to curb the disease.

Meanwhile, a parasitologist in Cairo called Arthur Looss was self-experimenting by swallowing an entirely different intestinal parasite. While eagerly searching through his stools he was surprised to find hookworm eggs. Having never socialised with miners, how had he acquired their worms? A few weeks earlier he had been giving hookworm eggs to experimental guinea pigs and a drop of the worm solution had fallen onto his hand. He kept his hands well away from his mouth and washed them thoroughly, yet he had become infected. To discover how, he deliberately put a solution of hookworms on the back of his hand. Within a minute or two the water on his hand was clear; hundreds of worms had vanished into his skin.

Looss made several unsuccessful attempts to clear the infection with thymol, a powerful disinfectant that irritates the kidneys. Some of the remedies on offer were worse than having worms. Thymol would be superseded by carbon tetrachloride, or dry-cleaning fluid as we now call it. Even in tiny doses it causes mental confusion, nausea and vomiting, followed later by necrosis of the liver leading to convulsions, respiratory failure and even death. Unfortunately for Looss, a safe and effective cure for hookworm was still twenty-seven years in the future.

Looss had initially become infected by chance, and now serendipity intervened again. A patient in a local hospital had to have his leg amputated, which was unfortunate for the patient but lucky for Looss. Immediately before the operation Looss splashed the leg with hookworm solution and later examined the detached limb. Dissection revealed that the worms had entered via the hair follicles, from which, given time, they would have migrated to the intestine.

Miners labouring underground often went barefoot, inviting infection from the millions of worms thriving in the damp dust underfoot. Merely getting them to wear shoes or even painting the soles of their feet with tar greatly reduced the rates of infection.

Hookworm is not confined to miners. In the Third World the majority of the population may go barefooted. It still infects 900 million people who have their energy sapped. Even in the southern states of the USA, the rural poor have long suffered from hookworm and, despite large eradication programmes, in some areas it is still said to infect fifteen per cent of children, retarding both their physical and mental development.

The tropics offer an even richer menu of parasites. In the nineteenth century the majority of explorers, soldiers and administrators who went to tropical West Africa never returned. Eighty-five per cent died or were 'reduced to mental and physical wretchedness'. India was almost equally dangerous. For two months of the year a quarter of the entire population was incapacitated. The culprit was malaria.

Often the smaller the parasite, the worse its effects. Malaria is caused by a microscopic organism called *Plasmodium*. By 1898 the basic life history of the parasite was understood and the disease was known to be spread by particular types of mosquito. Unlike many internal parasites that produce eggs to be expelled, *Plasmodium* multiplies itself within the blood stream and the number of retained parasites increases exponentially. The blood becomes the parasites' broth. They invade the red blood cells and if the cells are destroyed faster than the body can replenish them the consequences are dire.

Even if the victim's defences keep the infection in check some of the parasites sequester themselves in the body only to emerge again periodically, causing a relapse.

What was needed was a vaccine, but it was not until 1971 that a promising idea for one came along. David Clyde spent years in the tropics before becoming a researcher at the University of Maryland. He thought that irradiating malaria-carrying mosquitoes with just enough X-rays to weaken the parasites without killing the mosquito might be the first step towards producing a vaccine. The plan was to inject the now-benign parasites into a human and stimulate the body to produce antibodies against malaria.

For the tests, the hypodermics were mosquitoes and the guinea pigs were volunteer prisoners from penitentiaries. Clyde also became a volunteer because he thought it was essential to know exactly how a patient felt, and he would be able to describe any side effects far more accurately than the average convict.

He did far more experiments on himself in order to test whether the procedure would protect against different types of malaria. He became feverish and shivered uncontrollably, as well as vomiting. The bouts of fever returned every twelve hours until he cured the infection with drugs. He also ran tests to determine which strain of malaria produced the best vaccine. The technique was to fasten several cages to his arms, each containing several hundred infected but irradiated mosquitoes. He was bitten thousands of times which was very unpleasant. The next step was to see how effective the immunisation was by first being vaccinated, and then bitten by malarial mosquitoes. It worked: he didn't develop malaria.

In 1986 David Clyde was honoured by the World Health Organisation for his efforts, but sadly the vaccine that relied on irradiated mosquitoes could not be scaled up sufficiently for mass inoculation. Subsequent attempts at producing a vaccine have also been disappointing. *Plasmodium* is a formidable opponent − it has an immense capacity to mutate when challenged by a new drug, and can hide in the body away from harm.

Until the late nineteenth century, outbreaks of ague (mostly malaria) were commonplace in Britain and the disease was endemic in the East Anglian marshes of southeast England. We still have several species of mosquito capable of transmitting malaria should they become infected, and exotic mosquitoes are now appearing in Europe. With global warming the climate is likely to attract such immigrant insects, but the scarcity of people and animals with malaria in Britain, and the prompt treatment of those who contract it, should keep us safe.

In the 1950s we thought that malaria could be wiped from the face of the Earth. Global deaths from the disease fell by ninety-five per cent in the wake of extensive mosquito-eradication programmes, and a third of the world's population, living in previously malaria-stricken regions, are now free of the disease. But with the ban on the use of DDT for spraying inside houses malaria re-emerged. The mosquitoes soon became immune to other pesticides and in 2006 the World Health Organisation recommended the renewed use of DDT, despite concerns over its environmental effects. A billion-dollar grant from the health foundation of Bill and Melinda Gates has stimulated new initiatives, and a vaccine is in the pipeline. The results of the preliminary trials are promising. In the meantime malaria continues to kill a million people every year. Half the victims are children.

Not all parasites are the darkest of villains – some may even be beneficial. In the 1930s and 1940s patients suffering from syphilis-induced lunacy were deliberately given malaria. The infection was controlled with quinine and of 600 cases the mental state of almost half of them improved.

Parasites have cohabited with humans for millennia. Physiologically, we have usually come to a mutual understanding that allows both to thrive. Often infections in early life prepare the body's immune system for future encounters. In developed countries, children raised in clean houses with no pets suffer more allergies and asthma than those who are brought up in less fastidious conditions. Even a parasite or two may be preferable. Twenty-four patients suffering with multiple sclerosis, half of whom had recently been diagnosed as having parasitic worms, were studied for five years. In those with worms the disease progressed more slowly and relapses were far less frequent than in the parasite-free group.. Much larger-scale studies of Africans have indicated that people suffering with schistosomiasis rarely suffered from diabetes, rheumatoid arthritis or multiple sclerosis. These diseases, especially diabetes, are common and are increasing in developed countries. They are all autoimmune conditions in which the body's defences attack its own tissues and organs.

Invading parasites are recognised as foreign bodies by regulatory immune cells, the 'office managers' of the body's defence team. Their job is to marshal the host's immune responses to repel boarders. Humans have always been parasitised, so the regulatory immune cells have been kept busy. Perhaps now, with our 'old companions' eliminated, the under-employed immune cells may go into overdrive and cause autoimmune diseases.

On the other hand, parasites may enable us to control some conditions. In the laboratory, extracts of blood flukes have enabled diabetes-prone mice to stay free of the disease. It raises the possibility of a similar drug for people, to combat the scourge of diabetes. A trial with patients suffering from inflammatory bowel disease indicated that regular dosing with parasitic whipworms eliminated the symptoms.

Inflammation and allergies are usually triggered by an excessive immune response. Whipworms, flukes and hookworms survive within us by damping down our immune response that should attack them. John Turton, working at a medical research laboratory in England, relieved his hay fever for two summers by deliberately infecting himself with hookworms. When he got rid of the parasites his allergy returned.

Summer-holiday preparations may never be the same again: suntan lotion, insect repellent, dose of hookworms . . .

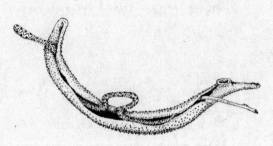

A male blood fluke lovingly enfolds an unsuspecting female.

The Desire for Disease

> 'Physicians of the utmost fame
> Were called at once, but when they came
> They answered, as they took their fees,
> There is no cure for this disease'
>
> — Hilaire Belloc

In the early nineteenth century it was inadvisable to be downwind of a city, for its streets were paved not with gold but with 'malignant effluvia'. Every day twenty-three tonnes of horse droppings were removed from London's two most fashionable shopping streets. Elsewhere, no one bothered. The fate of garbage was to accumulate and rot. There was no system for removing waste. In Leeds, for example, half the streets had no drains. In one area of the city there were only two privvies for 400 people. The local authority removed the accumulated effluent from just thirty houses — it required seventy cartloads.

Festering slums were breeding grounds for diseases. Edwin Chadwick's monumental report into the 'sanitary conditions' of the working class in 1842 revealed that labourers and mechanics in Manchester had a life expectancy twenty years shorter than rural labourers in the surrounding countryside. Fifty-seven per cent of the children of Manchester's poor never saw their fifth birthday. The gentry also fared better. They could expect to live well into their forties, whereas a labourer was lucky to see his twentieth birthday. As a Tory politician made clear, the lot of the poor 'must always be eating, drinking, working and dying'. But if the workers became diseased and died prematurely, their families were flung onto the street and became reliant on the state. Worse still, how could the well-to-do avoid catching the dreadful diseases intended for the poor?

Medics had little knowledge of how diseases were transmitted from one person to another. Probably it was through personal contact with the sick or perhaps from their bed linen or utensils. It was easy for the wealthy mill owner to avoid sharing the slum dweller's spoon, but there was a more pernicious mechanism at work.

The medical profession firmly believed that decay did not merely produce a bad odour but also airborne poisons — malicious 'miasmas' carried by the wind. Malaria, for example, means 'bad air'. Since 'all smell is disease' courtrooms were crammed with bowls of flowers whose scent would nullify the dangerous stench of prisoners.

The Houses of Parliament sat beside the fetid river Thames and the 'Great Stink' of 1858 shut down the seat of government to spare the 'nasally tortured' members. Perhaps from fear of being 'miasmatised' the House voted substantial funds to begin the construction of a network of sewers to carry the city's waste down to the lower reaches of the river.

This was undoubtedly the largest single contribution to London's future health, but it was not the smell of the river that had to be feared. Should you fall from London Bridge, it would be far better to drown quickly than to be rescued to die slowly from some dreadful disease. Yet the majority of citizens got their drinking water from the river.

A worldwide epidemic of cholera began in India in 1817 and lasted for almost twenty years, leaving mountains of corpses in its wake: forty million died in India alone; one in every twenty Russians perished. A succession of pandemics followed with high death tolls in Hungary, Japan and the United States. Three thousand pilgrims to Mecca died in a single night. In Britain there wasn't a city that escaped. Prompt action to deal with outbreaks was often impeded by merchants and other 'interested parties' who vociferously denied that cholera had arrived in their city. Medical authorities condemned people of 'gross and unexampled ignorance or shameful venality who . . . continue to deny the existence of an unusual disease, until the choked graveyards bear witness to the deplorable fact'. Even some doctors were accused of 'perverting and concealing facts which militate against their respective theories'. Perhaps 'when desolation and death thicken around . . . reason becomes silent', or the unreasonable become deaf.

Cholera strikes like a hammer. A seemingly healthy person can collapse and be swamped in vomit and pints of diarrhoea. One woman was stricken so suddenly that she fell into the fire. In Paris cholera struck a society ball with couple after couple collapsing mid-waltz. It was said that some were hastily buried, still in their evening gowns.

The lucky ones died within hours. Many of those who clung to life turned blue or black and succumbed to extreme dehydration. When they had no more fluid to void, they expelled fragments from the lining of their gut. The pain was 'like a sword put in on one side of the waist and drawn through, handle and all'. It is a ghastly way to die.

In 1832 a Dr Latta showed that patients with cholera could be saved by 'copious injections of saline fluids into the veins'. Sadly, other physicians continued to do the opposite, giving purges and vomit-inducers to patients who were expelling vast quantities of water from every aperture, presumably 'to assist in hurrying them to the grave'.

Countries were powerless to halt the disease's spread. Troops were stationed on their borders to keep out possible carriers. Ports quarantined ships from infected areas. At the beginning of the voyage of the *Beagle*, Darwin was prevented from landing in Tenerife 'by fears of our bringing the cholera'.

Diverse 'cures' and preventative measures were tried. Medics had great faith in 'distracting' the disease by applying severe irritants to the patient's skin. One remedy involved blowing hot air under the bedclothes – to such good effect that it set fire to the bed. Patients were given a 'flannel cummerbund' or swathed in bandages like a mummy and told to inhale toxic fumes of mercury. They should also drink 'tonics' containing ammonia and nitric acid or strychnine, and eat beef and peas boiled in chlorine.

There were 'official' days of fasting and humiliation to beg God to erase the pestilence. But if cholera was indeed a punishment for our sins, it was strange that during cholera epidemics prisons were the safest places in the land. Others who benefited were healthy tramps who turned up on the doorsteps of grand houses to announce that they were oozing with cholera and demand money to go away.

The greatest beneficiary was a London clergyman who established a life insurance scheme during a cholera outbreak in 1877. He gambled that most people didn't catch the disease and many of those that did survived. The premiums were only a penny a week, but he later retired a rich man when the Prudential had become a million-pound business.

Attempts were made to 'clear the air' of miasmas: towns were shrouded in smoke billowing from barrels of burning tar, citizens were deafened by the sound of mortars being fired heavenward every hour. It was even suggested that gunpowder should be exploded in bedrooms. These methods proved as effective as Joseph Addison's country gent who nailed shut his gate to keep out the crows.

Although the Board of Health favoured the miasma theory, a puzzling feature of cholera was that it didn't fit the airborne scenario. Sometimes among people living in close proximity and breathing the same air some contracted the disease while others didn't. Also, since the gut was affected first, the causal agent was perhaps ingested rather than inhaled.

That was certainly the belief of John Snow, a doctor who had trained at the Hunterian School of Medicine in Soho. He was a successful physician and by means of experiments on animals and himself he had invented an inhaler that delivered a regulated dose of anaesthetic to the patient. By removing much of the uncertainty from anaesthesia he became the expert in the field. It was he who gave chloroform to Queen Victoria to ease her deliveries.

Snow had been in Sunderland during England's first cholera epidemic and became seriously involved with the disease when there was an isolated outbreak in London in 1849. Within ten days almost six hundred people died in a small area of Soho. Florence Nightingale, a young nurse at the local hospital, could do nothing but watch them die. Karl Marx, a local resident, described Soho as 'a choice district for cholera'.

In the first-ever epidemiological study Snow plotted where

every victim lived and discovered that all those who caught cholera had drunk water from the Broad Street pump. None of the seventy workers at the Broad Street brewery fell ill because the brewery had its own well. Free beer was also available to employees and we now know that beer is poisonous to the organism that causes cholera. The local workhouse also had its own well and only five of its 535 inmates died of cholera. Removing the handle from the pump accelerated the decline of the outbreak. Later, the water supply to the pump was found to be contaminated from a leaking cesspool in a cellar only a few yards away.

In a later study Snow discovered that the incidence of cholera was fourteen times higher in areas of the city where the drinking water came from the Southwark and Vauxhall Company compared with those communities supplied by the Lambeth Company. Both the companies got their water from the river, but the Southwark and Vauxhall obtained their supply *downstream* from a large sewage outfall, whereas the Lambeth Company pumped up water from *above* the discharge. Dickens was right when he wrote: 'Look at the water. Smell it! That's wot we drinks. How do you like it?'

Although he couldn't name the causal organism, Snow had no doubt that cholera was a waterborne disease originating from contaminated drinking water. It did not convince everyone – in fact, it convinced hardly anyone. When he pleaded for the handle to be removed from the Broad Street pump, 'not a member of his own profession, not an individual in the parish believed that Snow was right'. Nor was there great enthusiasm for public health measures. *The Times* declared that as a nation 'we prefer to take our chance with cholera . . . than be bullied into health . . . There is nothing a man hates so much as being cleaned against his will, his pet dung heaps cleared away . . . It is a positive fact that many have died of a good washing . . . no longer protected by dirt.' A cleric railed against failing to give God credit for catastrophes and 'explaining away the Lord's visitation into a carnal matter of drains'. Snow died young in 1858 and it would be a generation before he was credited with his major discovery.

Diehard miasmatists were determined not to be persuaded. Even *The Lancet* published several articles dismissing Snow's conclusions. His sternest critic was a German chemist called Max Pettenkofer, whose achievements so far had been developing a method for separating platinum from gold, improving the quality of German cement and producing red Bavarian glass. On his appointment as Professor of Medical Chemistry at the University of Munich he turned his attention to improving public health and exploring the connection between disease and the environment. His idea was that the unhygienic conditions endured by the poor rendered them more susceptible to miasmas.

When cholera broke out in Munich in 1854 Pettenkofer mapped the geography of the victims, just as John Snow had done in London. He claimed that the majority of the deaths occurred in low-lying damp areas and he amplified this into a grand theory. Fresh air permeated into these damp soils and instigated an unspecified chemical reaction to produce a poisonous miasma. He didn't attempt to isolate or identify this deadly emanation.

Elsewhere others were finding a very different cause of disease. Louis Pasteur in Paris and Robert Koch in Berlin were independently developing what became known as the germ theory of disease. Pasteur demonstrated that decay was not merely a chemical reaction; it was mediated by microbes and could be arrested by pasteurisation (repeated heating and cooling). He also developed a vaccine for rabies.

Koch, his bitter rival, developed microscopes that allowed him to see and characterise these tiny bacteria. He was a brilliant microbial detective. He and his team tracked down the microbes responsible for anthrax, gonorrhoea, leprosy, pneumonia, typhoid and syphilis. He also produced a set of rules by which the causal agent of a disease could be established:

1. Isolate the bacterium from an infected animal.
2. Culture it in the laboratory. For this job Koch's assistant Julius Petri invented the Petri dish.
3. Demonstrate that the microbe produces the symptoms of the disease when injected into a healthy animal.
4. Finally, re-isolate the microbe from the infected animal.

The problem with cholera was that it didn't infect animals, only people, so it was difficult to experiment with. Both Pasteur and Koch sent teams to Egypt during an epidemic to study the disease. The co-leader of Pasteur's team died of cholera.

When the Egyptian outbreak waned Koch went on to India where cholera was always available. He isolated from both live and dead victims a tiny comma-shaped bacterium that he called *Vibrio cholerae*. He had found the cause of cholera.

Max Pettenkofer was not impressed. He scoffed at those who 'confine themselves to the behaviour of the comma bacteria in test tube or plate'. He conceded that the bacterium *might* play some role, but the disease could only develop if certain environmental factors were favourable such as the 'right' soil — no matter that several researchers had examined the soil where outbreaks were rife and found that it wasn't the 'right' type. Koch also showed that the cholera bacterium was present in the water supply drunk by victims. And in Hamburg, where either side of the same street had a different water supplier, only those on the contaminated side fell ill. It was, as Snow had asserted, a waterborne disease.

When Koch referred to him disparagingly as 'Herr Localist', Pettenkofer could take no more. In 1893 he requested a sample of *Vibrio cholerae* and Koch's laboratory obliged. Pettenkofer had no intention of studying the bacteria; he had a far more dramatic experiment in mind.

Propelled by the certainty that he was right, he would put Koch in his place.

When Pettenkofer had been young he had had theatrical ambitions. In Goethe's *Egmont* he had played the unsuccessful lover who feigns to swallow poison but wisely refrains. This time the seventy-four-year-old professor in love with his theory held the flask of bacteria aloft and announced to his assembled colleagues: 'Even if I deceived myself and the experiment endangered my life, I should face death calmly, for it would not be a thoughtless or cowardly suicide. I should die in the cause of science.' To the horror of his audience he drank the contents of the flask.

He suffered severe stomach cramps and diarrhoea that lasted for a week. These were symptoms of cholera but relatively mild. How did he cheat death? Did a concerned colleague heat the culture to make it less virulent or was it a less dangerous strain, or was he just lucky? In epidemics many people contract the disease but don't die.

Pettenkofer sent a note to Koch boasting that: 'Herr Doctor Pettenkofer has now drunk the entire contents . . . and is happy to be able to assure Herr Doctor Professor Koch that he remains in his usual good health.'

He was convinced he had disproved 'once and for all' the idea that drinking water could spread cholera. He rejected all this 'zealous comma-hunting' for without the 'cholera miasm' from the soil the bacterium was harmless. But he was swimming against a rising tide. His theories were progressively discredited by everyone except his former students. Even fellow professors in Germany dismissed them as 'nonsensical' or 'the product of an inventive imagination . . . built on drastic hypotheses that are entirely contradicted by the real facts'.

Pettenkofer's case highlights the danger inherent in medical research reliant on trying to find correlations between external factors and specific diseases. Because in the incident that he investigated cholera was more prevalent in damp places, Pettenkofer leapt to the conclusion that damp soil was implicated in causing the disease and then devised a theory to explain how. The problem is that correlations frequently occur between totally unrelated factors. For example, the number of pigs reared in the United States followed the same trends as the production of pig iron. This did not necessarily indicate that pigs provided the raw material for ingots. In Britain there was found to be a correlation between the number of TV licences bought and the incidence of mental illness. Was this proof that TV addled the brain or that only the mentally impaired bought licences? Or was it just a chance correlation of unrelated things? The news is full of reports such as that natives of Tuscany use more olive oil and drink more red wine than the British, and have a lower incidence of heart disease. So should we get oily and drunk to prevent heart attacks? There may be dozens of reasons why Tuscans suffer less from cardiac problems: unlike the Brits they don't consume two million tonnes of chips every year, or think that taking a deep breath is a form of exercise. One could argue that a heart attack isn't such a bad way to go compared to the horrors that may lie in wait should your heart keep going.

For his undoubted achievements in the field of public health Pettenkofer was rightly rewarded with a gold medal from the British Institute of Public Health. In Germany he was elevated to the aristocracy as His Excellency Max *von* Pettenkofer. But even these honours failed to quench his bitterness as the support for his theory of infection proved to be as insubstantial as a miasma.

Nine years after he had risked his life by tasting cholera, His Excellency conducted his final experiment by putting a revolver to his temple and pulling the trigger.

Had Pettenkofer sacrificed himself to cholera he would not have been alone among self-experimenters. Curiosity has killed more than cats.

In 1885 Daniel Carrión, a Peruvian medical student who was studying a skin disease called verruga, inoculated himself with blood from a patient's 'wart'. He was soon dangerously ill and it became clear that he was suffering from a fatal blood disease called Oroya fever. Daniel realised his plight, as a friend had recently died of the same disease, but he also appreciated what he had discovered. 'This is the evident proof,' he said, 'that Oroya fever and the verruga have the same origin.' Within weeks twenty-six-year-old Daniel was dead and the physician who had initially discouraged him from experimenting but had then assisted him was arraigned for murder, although he was later acquitted. Daniel was acclaimed in Peru and there is a statue of him in Lima.

The search to determine how yellow fever was transmitted was also not without casualties. As with so many epidemic diseases, yellow fever influenced history. Had Napoleon's Caribbean army not been ravaged by the fever, which prompted the sale of Louisiana and a vast amount of other land, North America might well have become French.

Yellow fever damages the liver, causing jaundice. The blood fails to clot and leaks into the stomach to be regurgitated as black vomit. The illness is often fatal within days. Clearly not a disease to messed with, but that didn't stop the researchers. Although there was no evidence that carers caught the fever from their patients, often relatives merely abandoned sufferers for fear of contracting the disease. In 1804 Stubbins Ffirth, an American medical student, decided to determine that it was contagious. He slept alongside fever patients who were 'attended by black vomit' and ensured that he received 'the breath of patients in my face'. But he had only just begun to explore the encyclopaedia of unpleasantness.

Black vomit may not be everyone's cup of tea, but Ffirth spent hours inhaling the vapour from simmering vomit until he felt too nauseous and faint to continue. Although a dog he had injected with black vomit died within minutes, he injected it into his own veins and into deep cuts on his arms. He smeared his body with blood, sweat and urine, and drank the saliva, blood and vomit of infected patients. As someone who declines a second helping of black pudding, Ffirth's appetite for black vomit wins my admiration.

He didn't catch the disease, and therefore wrote a reassuring article: 'I hope these experiments will have a tendency to allay . . . that great fear which some have . . . as it is at least doubtful whether it is ever communicated from one person to another by means of contagion.'

After all Ffirth's ordeals, the publication of his results had little influence for he had not shown what *caused* the disease. Almost a century later the cause was still unknown and there was concern over the high death rate of troops in the Spanish–American War. An American army medical team led by Walter Reed was dispatched to Cuba to discover how the disease was transmitted.

Mosquitoes were suspect as they were being implicated in the transmission of malaria and filariasis (elephantiasis). Yellow fever doesn't occur in animals so there was no choice but to experiment on humans. In Cuba, where yellow fever was endemic, it would be almost impossible to find many non-infected volunteers and urgency was in the air as experts from the Liverpool School of Tropical Medicine had arrived on a similar mission. The team also had 'a grave sense of responsibility . . . which the conscientious observer must always feel, even with the full consent of the subjects to be experimented upon'. So they made a pact that they would all consent to experiment on themselves. No sooner was this agreed than Reed, the team leader, departed for Washington.

The procedure was simple. Mosquitoes were placed on the arms of patients with yellow fever and were then used to infect the experimenter. The first to submit to this was a bacteriologist called Jesse Lazear, but he didn't develop any symptoms. Next in line was an Englishman, James Carroll, who had studied medicine in the United States. He had a wife and four children. Earlier he had been a guinea pig in tests of an experimental vaccine against typhoid, but by accident the inoculum still contained live bacteria and seven out of the twelve volunteers went down with typhoid. Carroll had avoided infection, but this time he was not so fortunate. Within days he was seriously ill with yellow fever and his appearance shocked his colleagues. His yellowed eyes were bloodshot, and he became feverish and too weak to stand.

Although Carroll's life had been in danger, Lazear, who was later described as being 'oblivious of self', was exposed again to infected mosquitoes to confirm the findings, which he did with a vengeance. He was soon spurting black vomit, and there was fear in his eyes before he became delirious. Within twelve days he was dead. He was thirty-four years old. His pregnant wife received a telegram that merely stated: 'Dr Lazear died at eight p.m. this evening.' She hadn't even been told that he was ill.

The report of their findings was published in record time, only two months after Lazear's death. Reed wrote: 'The mosquito serves as the intermediate host for the parasite of yellow fever, and it is highly likely that the disease is only propagated through the bite of this insect.' The experts weren't convinced and the press were scornful. According to the *Washington Post*: 'Of all the silly and nonsensical rigmarole about the yellow fever that has yet found its way into print . . . the silliest beyond compare is to be found in the arguments and theories engendered by the mosquito hypothesis.'

So the experiments continued, this time with local volunteers, each of whom was given $100 in gold, with an extra $100 if they caught the fever. Those rejected wept with disappointment. Some volunteers had to sleep for three weeks in the pyjamas and bed sheets soiled by yellow fever patients and with their heads resting on pillows covered with a towel soaked in a patient's blood. None of the volunteers caught the fever, which merely replicated Ffirth's results from similar ordeals ninety-three years earlier.

To identify the causal organism volunteers were injected with infected blood that had been filtered to remove all bacteria. They still became feverish, because the microbe responsible was a virus (although no one knew this at the time), which is much smaller than a bacterium and could pass through the filter.

Others were injected with blood from a mildly infected

patient in the hope that this would act as a vaccine conferring some immunity. These injections killed a young nurse. Two soldiers also died and another became so incapacitated that only public donations saved this 'soldier who gave himself to science'. All volunteers had to sign a cunningly worded waiver indicating that he might develop a fever which 'endangers his life to a certain extent, but it being entirely impossible to avoid the infection during his stay in this island, he prefers to take the chance of catching it intentionally . . .'

Walter Myers, one of the Liverpool team, also died of yellow fever and Carroll never fully recovered his health. Sadly, Carroll and Lazear received little recognition for the risks they took, whereas Reed, the team leader who was a thousand miles away at the time of the fateful experiments, was lauded as a valiant pioneer and his name was immortalised in the Walter Reed Military Hospital in Washington. Perhaps the moral is that it is far better to write up the report than to take part in the experiments.

Yellow fever was not eradicated from the United States until 1905 and remains a serious threat in the tropics. The causal virus has now been identified, but once the disease has taken hold there is no cure. Mass vaccination is effective, but in 2008 a shortage of vaccine compromised the campaign worldwide.

Even when a vaccine is developed it may not be well received. A British Army doctor called Almroth Wright developed a vaccine against typhoid fever. Existing vaccines involved giving the patient a mild infection of typhoid to stimulate the body's defences. Unfortunately, sometimes the 'mild' typhoid killed the patient. Wright's idea was to inject dead typhoid bacteria in the hope of fooling the immune system into thinking it was under attack, thus producing immunity with no risk of getting infected by the inoculation. The only problem was that to test whether or not it was effective, the body had to be challenged with the real disease. So in 1897 Wright and his laboratory assistants injected themselves with the vaccine and then with typhoid. It worked.

The new vaccine was issued to British troops departing to fight in the Boer War. Such was their suspicion of the injections that boxes of vaccine were flung into the sea. As a consequence very few of the soldiers benefited from the vaccine and 9,000 died from typhoid. Even today 600,000 people die every year for want of a vaccination.

Prejudice against new findings that challenge existing beliefs is not a thing of the distant past. When I was young it seemed to me that not only my father but all my mates' fathers suffered from duodenal ulcers. Every doctor knew they were caused by stress, smoking, bad diet, alcohol. The treatment was to take antacids or have major surgery.

In the early 1980s Barry Marshall, an Australian microbiologist, teamed up with a pathologist called Robin Warren to document the bacteria living in the human gut. They found that one species, *Helicobacter pylori,* was invariably present in patients suffering from duodenal ulcers and also in seventy-five per cent of those with stomach ulcers. Could it be the causative organism?

To find out, Marshall slid a tube down his throat into his stomach and removed fragments of the stomach lining. These were examined to ensure that he didn't have either a gut infection or *Helicobacter*. After allowing time to let the gut wall heal he swallowed a culture of the bacterium. He had, of course, taken precautions before doing so. Firstly, he didn't inform the hospital ethics committee in case it refused permission and secondly, he didn't tell his wife until after he had taken the draught. She guessed anyhow when within days he became listless and began vomiting. To add insult to self-inflicted injury, his wife informed him that his breath was 'putrid'. A series of biopsies of his gut tissue revealed severe inflammation (gastritis) that precedes ulceration. Fortunately he cleared up the problem with an antibiotic. Marshall and Warren went on to show that when the *Helicobacter* bacteria were eliminated, symptoms of gastric ulcers vanished within days and the ulcers began to heal even in patients who had suffered from them for decades. Yet it was thirteen years after Marshall's experiment before this treatment became widely available in western hospitals. During that time perhaps hundreds of thousands of patients were given the wrong drugs or underwent unnecessary surgery.

At first critics scoffed at the idea that gastric ulcers were an infection since everyone knew they were the result of a chemical imbalance. As for bacteria living in the stomach, it was far too acid an environment for that. Worst of all, perhaps, Marshall was a junior doctor at the time and neither he nor Warren was a gastroenterologist. What did he know about such matters?

Well, sufficient to be awarded the Nobel Prize for medicine in 2005.

A mosquito caught in the act of transmitting one disease or another.

The Disease Detectives

'*Infectious disease is one of the few genuine adventures left in the world*'
– Hans Zinsser, who discovered the causal organism of typhus

When a deadly disease threatens, our instinct is to flee. Surely only fools rush in. Yet several countries have specially trained teams of medics, scientists and vets on standby to investigate outbreaks of disease wherever they occur. The Epidemic Intelligence Service of the United States Centres for Disease Control is just one example of these storm troopers of disease. The epidemics that interest them most are also the most deadly.

In 1967 three patients in the clinic of the University of Marburg in Germany developed alarming symptoms. As if fever, vomiting, diarrhoea and severe pain were not enough, blood began to suffuse their entire body beneath the skin. Then blood leaked from every opening, even their eye sockets.

They were bleeding to death and no one had the slightest idea what had struck them.

The clinic soon had seventeen more patients with the same symptoms. All were dangerously ill or dying. The illness was highly contagious and a doctor and nurse became infected.

Microbiologists worked night and day to identify the causal organism, which proved to be a virus. It was a previously unknown disease so they called it Marburg fever.

In Marburg and other haemorrhagic (blood-flowing) fevers small clots form all over the body cutting off the blood supply to tissues downstream, which then die from lack of oxygen. The body responds to this critical problem by flooding the system with anti-clotting compounds. This emergency reaction results in massive internal bleeding. Several other conditions are dangerous because of an excessive response to invaders by the body's defences.

The three original victims worked for a local pharmaceutical firm. They had all been in contact with live monkeys whose liver cells were used to produce polio vaccines. The monkeys, imported from Uganda, had brought the Marburg virus with them and it had jumped from monkey to man.

There was worse to come. In 1976 outbreaks of haemorrhagic fever occurred in the Sudan and at a mission hospital in Zaire. The virus was similar to the Marburg one, but not identical. For the 318 victims in Zaire the mortality rate was ninety per cent, well over three times higher than for

Marburg. They named this new disease Ebola after a tributary of the river Congo. Had even one of these patients travelled abroad it could have sparked a deadly pandemic. There was no cure; indeed, there was no treatment at all.

Disease detectives were dispatched to the outbreaks. The impact of the intrusion of western scientists can be as frightening to tribal communities as the disease itself. Imagine people wearing 'spacesuits' and perhaps military respirators, so as not to breathe the possibly contaminated air, walking down your street and taking throat swabs and samples of your blood.

There is a tendency for westerners to dismiss tribal medicine, with its witch doctors and belief in evil spirits, but in several African countries when the tribal elders realised this was no ordinary infection they enforced appropriate emergency measures. The infected were confined to their huts and tended by only one person. When victims died their huts were torched and their bodies were buried well beyond the village boundary and without a funeral gathering. Those who recovered remained in isolation for a lunar cycle. No one was allowed to travel to another village. Such wise measures were often sufficient to prevent the spread of even the most contagious diseases.

Haemorrhagic fevers are the deadliest diseases on the planet. They should be handled under what is called 'hotzone' biosafety Level 4. This means that the researcher must be housed in a secure laboratory where all airflow is *into* the lab. Body fluids that may be infected must be 'handled' in special cabinets with the researcher wearing gauntlets, having sealed protective clothing covering the entire body, and breathing from an external air supply.

The conditions when researchers arrive in a remote jungle village are dangerously different. They may be uncertain which disease they are dealing with, and if it is not clear how it's transmitted it is difficult to know how to avoid it.

When processing hundreds of blood samples, mishaps are inevitable. During the outbreak of Ebola in the Sudan, Joe McCormick from the US Centres for Disease Control was taking numerous blood samples from patients. He accidentally stabbed his hand with a needle containing blood from someone with probable symptoms of Ebola. Though shocked and apprehensive, Joe reckoned that if he'd become infected he could do nothing about it, so he bravely continued to take samples. Luckily he was all right because the feverish patient did not have Ebola. There is a long tradition of researchers contracting the diseases they study. Joseph Goldberger, a pioneering disease detective in the early twentieth century, caught dengue fever and yellow fever and almost died of typhus. He also got a skin infection called Shamberg's disease, but that doesn't count because he gave it to himself deliberately in a self-experiment. We will hear more of his exploits later.

Goldberger was lucky to survive. Others were less fortunate. In 1927 alone three senior specialists in tropical diseases died of yellow fever while in the field. Deaths became so common that one scientific journal issued lists of the latest 'martyrs to medicine'. Even today, every team that flies out to confront outbreaks of deadly diseases knows of a colleague who has not survived. Yet C. J. Peters, a veteran disease detective, grumbles not about the risks of infection but about the bone-hard biscuits, cheese that tastes like putty and a foul paste of congealed fat masquerading as meat. He also fears foreign toilet paper that is best suited for sanding down furniture. His golden rule is to take his own loo rolls. He reckons that if you've put your ass on the line, you should pamper it a little.

Outbreaks of Ebola and Marburg occur sporadically and then 'go into hiding'. Discovering which animals act as reservoirs of the virus may help to devise ways of controlling the disease. With exposure to the disease over many generations wild animals may develop some immunity and can harbour the virus without displaying symptoms. Taking blood samples from such creatures is risky. Infection may be just a scratch or a bite away.

The 'reservoir' for Marburg was elusive until tourists visiting caves in Uganda died of the disease. At no small risk to themselves scientists caught every type of animal found in the caves and eventually discovered the Marburg virus in cave-dwelling bats. Later, other researchers showed that forest-living bats harboured Ebola.

For all the courage of the disease detectives, the unsung

martyrs of lethal outbreaks are the carers who tend to the sick and dying. Of the 774 people who died during the well-publicised SARS epidemic in 2002, 162 were hospital staff who bravely stayed at their posts. During the first outbreak of Ebola in Uganda nurses working at the centre of the epidemic threatened to strike until Dr Matthew Lukwiya told them that: 'Whoever wants to leave can leave. As for me, I will not betray my profession. Even if I am on the wards alone, I will continue.' He knew the risks and six weeks later he became one of the 224 victims of the outbreak. In 2007 Uganda was so terror-struck by another Ebola epidemic that the country's president advised people not to shake hands. Market vendors took to wearing gloves and priests refrained from handing out communion wafers. Again, a valiant local doctor headed for the focus of the outbreak. His boss had warned Jonah Kule that it could be a fatal mission, but he replied that he was prepared to die to help the stricken people. Within a few weeks both he and the matron who nursed him were buried.

Isolated facilities for the study of lethal diseases were established by the opening of the twentieth century when Alexander Oldenburgsky, a Romanov prince, set up a laboratory in an abandoned fort on an island in the Gulf of Finland. Its remit was to study plague and develop a vaccine. Few visitors were allowed and none could stay overnight. The researchers wore protective clothing such as rubber-lined cloaks, and were aware of how dangerous their work could be. There was the constant reminder of the isolation block. If you were contaminated and passed through its heavy hermetically sealed doors, you were unlikely to return. Two scientists died after being accidentally infected. The incineration of their bodies contributed to the in-house heating.

One of the staff, a physician and epidemiologist called Ippolit Deminsky, wanted to disprove the xenophobic notion that all plague was imported into Russia by foreigners. So he trudged the steppes searching for proof that plague lurked in the local fauna. He isolated the causal bacterium from a dead rodent (a suslik), but infected himself in the process. In a telegram to his colleagues, he instructed them to 'take the cultures that I have isolated. All the laboratory records are in order ... my body should be examined as an experimental case of a human contracting the plague from suslik. Goodbye.'

By 1981 the fort had been replaced by over a hundred laboratories ostensibly dedicated to the development and manufacture of vaccines. But they had another secret and sinister purpose.

That deadly diseases might be a weapon of war was a very early idea. In 1346 the Mogul hordes besieging Kaffa on the Black Sea coast hurled corpses riddled with plague over the city walls. The citizens who survived fled west taking the plague with them and, it is claimed, instigating the Black Death, the greatest catastrophe in human history. Once deployed, biological weapons are difficult to control.

Anthrax has long been the killer of choice for makers of biological weapons. It is a highly infectious bacterial disease of livestock that can transfer to humans. Virgil described the horrors of anthrax: 'In droves she deals out death and in the very stalls piles up the bodies, rotting with putrid foulness, till . . . men bury them in pits.' He also noted that if a person wore the fleece from an infected beast, he suffered 'feverish blisters and accursed fire feeding on his stricken limbs'. In later centuries it came to be called the 'woolsorters' disease' and once laid low a hundred felt workers in a carpet factory.

In the years preceding the Second World War several countries were working feverishly on the production of biological weapons. The Japanese produced bombs made of porcelain, which could be shattered by a tiny charge to scatter their payload of anthrax spores without damaging them. They tested them on foreign prisoners staked out in batches of ten. Germ warfare presented an excellent opportunity to experiment on fellow human beings on a large scale and anthrax became one of the first weapons of mass destruction. During their Manchurian campaign against the Chinese the Japanese deployed their anthrax bombs, infecting entire towns and villages. Unfortunately, a mischievous wind changed direction and over 10,000 Japanese troops became ill and almost 2,000 died. The Japanese also gave contaminated chocolate to prisoners of war, then allowed them to return home taking the disease with them.

Had Germany used biological weapons, the British were

ready to retaliate with five million cattle cakes produced in a disused soap factory and then laced with anthrax spores by the housewives of Salisbury. Twelve Lancaster bombers would have dropped them over Germany. Churchill was enthusiastic about using anthrax to kill livestock and plague to poison armies.

There is a more deadly form of anthrax than over-seasoned cattle cake. Handling infected animals or their pelts transmits cutaneous anthrax with nasty lesions on the skin and a mortality rate of only twenty per cent; eating infected meat that's been cattle-caked does better with fifty per cent fatalities, but *inhaling* anthrax spores results in an aggressive wholebody infection that can kill ninety per cent of victims. So the wartime research concentrated on delivering anthrax by air.

In 1942 Paul Fildes, a bacteriologist who was the first head of Britain's Porton Down laboratory, calculated the amount of anthrax required for a strategic attack. As the laboratory was close to residential areas unsympathetic to anthrax assaults, the aerial tests were carried out on the island of Gruinard off the west coast of Scotland. Its only inhabitants were 155 sheep, but not for long – anthrax bombs saw to that. Fildes reckoned that weight for weight, anthrax could be as much as a thousand times more potent than any chemical weapon available at that time.

In 1943 the American biological warfare programme produced 7,000 bombs filled with anthrax. By the end of the war the United States had a factory that could produce 363 kilos of concentrated anthrax slurry with every run of its production line. It was never used.

Such was the lingering fear of germ warfare that after the war US researchers tested the dispersal patterns of harmless powder (standing in for anthrax spores) at Washington DC airport, in the New York subway and from a plane over the Bay area of San Francisco. In 1963 they played out similar aerial attack scenarios over three other cities.

Anthrax has long been considered a possible terrorist weapon. It's the poor man's 'dirty bomb' and, according to experts, it is not that difficult to produce. A 1993 report estimated that if a hundred-kilo cloud of its spores were released upwind of Washington DC it could cause anything up to five million deaths. Decontaminating the ground afterwards would be a mammoth task. The island of Gruinard remained under quarantine for fifty years with no one allowed to land. Viable spores of anthrax were still present and the land had to be decontaminated by removing loads of the most infected soil and spraying other areas with 280 tonnes of formaldehyde and 2,000 tonnes of seawater. The clean-up of tiny Gruinard took more than nine years.

America's fears were realised when in 2001 a letter containing white powder was opened by a newspaper photo editor in Florida. Four days later he died from inhaling anthrax. There had been only eighteen cases of inhalation anthrax in the entire United States in over a century, but within days more poisoned letters arrived on the desks of broadcasters and politicians. In Washington, postal workers at the main sorting office fell ill. Both the sorting office and the House of Representatives closed down.

Eighty investigators from the Centers for Disease Control (CDC) were soon on the scene. All the offices at NBC had to be quarantined and 'swept' for contamination. At the Senate Office, where the spores had been dispersed by the air-conditioning system, two entire floors were sealed off and checked. Over 400 people might have been infected. Everyone had a nasal swab that was then cultured to see if it contained anthrax bacteria.

The leader of one of the CDC teams was asked by a woman who had been feverish for two weeks whether, if she had anthrax, she'd be dead by now. The scientist reassured her that she would. Another man complained of his wife refusing to let him into the marital bed as he might have anthrax. The investigator suggested that she would have to think up a new excuse.

Twenty-two people were affected, some just from handling the envelopes and therefore getting the much less dangerous cutaneous anthrax. Five died, but no one was arrested.

It was almost seven years before the FBI homed in on the culprit. DNA taken from the fatalities showed that the strain of anthrax originated from the Army Medical Research Institute of Infectious Diseases at Fort Detrick in Maryland. The researcher who had created this strain and maintained it was Dr Bruce Ivins, who headed a group developing a vaccine against anthrax. It was alleged that he had sent 'defective' samples to the FBI when involved in the investigation of the anthrax-contaminated envelopes. A few days before the anthrax attacks, he sent an e-mail warning that Bin Laden's terrorists had access to anthrax. On being informed that he was to be prosecuted for the attacks, Ivins took a lethal dose of painkillers. His premature death ended the investigation and started the inevitable conspiracy theories. As one FBI agent remarked, 'There's always going to be a spore on the grassy knoll.'

The Institute at Fort Detrick is *the* centre for bioweapons research in the United States. It has the highest level of security. The regulations covering scientists working on lethal microbes state that they must be mentally and emotionally stable, physically competent and trustworthy. This bears little similarity to the description of Ivins given by the FBI. Allegedly, he had confessed to a friend that he was suffering from scary paranoia. His psychiatrist classified him as being a homicidal sociopath and his brother said he thought of himself as God. I once saw a sign on the door of a high-security research laboratory that said:

Caution! May contain nuts

Further insight into the effectiveness of the vetting procedure at Fort Detrick comes from the strange affair of Steven

Hatfill. He held high-security clearance at the Institute and had access to deadly strains of anthrax, Ebola and plague. After being designated a 'person of interest' in the anthrax attacks, though not prosecuted, he sued the US Department of Justice. His reward was $2.4 million up-front plus an annuity of $150,000 a year for twenty years.

A joint enquiry by the New York magazine *SEED* and the British newspaper *The Observer* revealed that Hatfill's immaculate CV was less than accurate. He had not graduated from the university he cited, nor was he a fellow of the Royal Society of Medicine, nor had he been a member of the SAS and NASA. His references from distinguished professors were faked. Clearly there are excellent career prospects for candidates with an inventive mind and imaginary qualifications. An interest in the deadliest germ on the planet is an advantage.

The United States alone has around 400 labs with over 15,000 people handling dangerous biological agents. At Fort Detrick four workers have died from accidental contamination, two of them from anthrax. Three are commemorated in the names of local streets.

One takes it for granted that such dangerous research activities are carefully monitored. Yet in 2007 and 2008 alone several of these institutions failed to ensure adequate safety precautions, and some even permitted unauthorised personnel to handle anthrax and Ebola. Others failed to report lab-acquired infections. Institutions have their own internal committees to assess safety in laboratories. But one committee had not even felt it necessary to meet when it approved the re-creation of the Spanish-flu virus that had once killed two per cent of the world's population. Researchers at a children's hospital in California were exposed to anthrax when a supplier accidentally sent live bacteria instead of the dead ones they had ordered. In 2005 another laboratory inadvertently sent out 4,000 virus-testing kits that included a virulent Asian-flu virus that had killed millions.

One would like to think that security at the 800 universities, institutes and commercial laboratories in the United Kingdom is better, but the revelation in 2008 that the firm responsible for security in government departments had not even vetted its own employees is not encouraging. MI5 and MI6 claim to have intercepted up to a hundred suspects posing as postgraduate students aiming to acquire skills in handling biological and chemical toxins. One pupil who graduated with a PhD in Britain was Rihab Taha, otherwise known as 'Dr Germ', the mother of Saddam Hussein's biological weapons programme.

Even in the best-run laboratories accidents occur. Anna Pabst died from meningitis when the animal she was injecting squirmed, causing the serum to squirt into her eye. Jeff Platt was injecting guinea pigs with Ebola virus when he accidentally pricked his thumb. He instantly removed his glove, bled his thumb and washed the wound with bleach. He became seriously ill, but prompt action saved his life. In

2009 a German researcher also accidentally stabbed herself with an Ebola-laden needle. She was given an experimental vaccine and survived.

Personal accidents are one level of risk, but institutional incidents are quite another. In the 1940s a joint US/Canadian biological weapons laboratory was beset with problems. Perhaps because of its remote location, there was low morale and creeping paranoia. The scientists became convinced that local flies were contaminating the canteen food with anthrax. The real problem with the Canadian lab was sloppy safety procedures and frequent contamination accidents. Fortunately the facility was closed before there was a major mishap, but some *have* occurred.

In 1979 there was an outbreak of anthrax in the Urals. Sverdlovsk (now with its original name of Yekaterinburg restored) was known as the place where Czar Nicholas II and his family were executed in 1918. Its factories forged the T-34 tanks that enabled Russia to repel the German invasion in 1943. It now became the focus of interest for epidemiologists abroad. Russian officials announced that the Sverdlovsk outbreak resulted from people eating infected meat, and stressed, perhaps a little too vociferously, that the incident had no bearing on the Soviet Union's compliance with the international convention banning bacterial and toxic weapons. Observers abroad were unconvinced, but it would be thirteen years before *glasnost* enabled an American team to visit Sverdlovsk to investigate the 1979 outbreak.

On their arrival a welcoming official advised them that there was 'no reason to disturb the skeleton in the cupboard'. The team were encouraged to hear that there was indeed a skeleton awaiting disturbance. If disaster strikes, the reflex response of governments is to lie. When a flood exhumed the skeletons of executed dissidents in another Russian town, they were dismissed as animal bones. When nobody swallowed that, they were said to be the remains of deserters. Similarly, when a dead sheep 'anthraxed' on Gruinard washed up on the Scottish mainland and infected a flock of sheep that then had to be slaughtered, the British authorities came up with an implausible tale of diseased sheep being heaved overboard from a Greek ship.

The investigation at Sverdlovsk was hampered because all the patients' records were destroyed by the KGB on orders from the Council of Ministers. The skeleton wasn't just rattling against the cupboard door, it was positively dancing to get out. One team of pathologists had bravely carried out autopsies on forty-two victims of the outbreak. Their only 'protection' was flimsy gauze mouth masks. Fortunately, they had retained tissue samples and anatomical photographs of each of the victims that revealed the classic symptoms of *inhalation* anthrax. But where could an aerosol of anthrax have possibly come from?

The American team plotted the exact location of every victim on the days immediately before the first people fell ill. Almost every one lay in a narrow zone running from north-west to south-east. Six villages where animals had died from anthrax at the time also lay along the same line. At the northern end of the zone was Military Compound 19. On the critical date the zone lay directly downwind of the compound. This indicated that a plume of anthrax spores had been liberated from somewhere in the compound. Five victims who'd lived and worked well outside the killing zone had the misfortune to be attending military reservist classes in the compound on the fateful day.

Beyond the ornate metal gates of Compound 19 there was a research laboratory ostensibly developing vaccines against deadly diseases, including anthrax. Testing such vaccines usually involved immunising animals with the vaccine, then 'challenging' them by exposure to a virulent strain of the disease. Such experiments would be carried out in a sealed chamber and after the experiment the contaminated air should have been expelled through highly efficient filters that removed the microbes. But the filters had to be well maintained.

Nine people in the compound died from anthrax and sixty more were killed in the city. Most died within three days of the first symptoms appearing. The surprisingly low number of fatalities might indicate that the amount of anthrax released was very small, but the prompt actions of the authorities probably greatly reduced the number of victims. Eighty per cent of the local population were vaccinated against anthrax and were also given a variety of antibiotics. Buildings were hosed down with disinfectants, hundreds of stray dogs were shot, infected livestock were killed and burnt, healthy animals were vaccinated. Human victims who died were hastily buried without ceremony in coffins filled with chlorinated lime.

Curiously, the military took no part in these operations. They never emerged from their barracks. Many of the relatives of victims had no doubt that the outbreak had something to do with what went on behind the walls of Compound 19, but the military never explained or said they were sorry. Indeed, a letter from an anonymous worker at the laboratory published in the newspaper *Izvestiya* described his research as being patriotic and dismissed the outbreak with the phrase 'accidents happen'.

In 1992 President Yeltsin, in a sober moment, blamed the military for the anthrax incident and the chairman of the KGB was told to scrap all germ warfare research at Compound 19. It was merely transferred to a new location. Yeltsin also decreed that all the relatives of the anthrax victims should be given a special state pension. It is claimed that no one received a single rouble.

That Unhealthy Glow

'What you can't see won't hurt you' — one of my mother's many truisms that were untrue

All scientists are detectives searching for clues to solve mysteries, questioning suspect information and probing the unknown. The unknown is often the unseen and mankind has long been fascinated by invisible forces.

The Renaissance physician Paracelsus was one of the first medical men to become obsessed with magnetism. He used a magnetised crystal to transfer disease from a patient to germinating seeds, and spawned generations of quack 'magnetisers'.

It was well known that sword wounds were best healed by a magnetised blade dipped in a mixture of the casualty's blood, two ounces of human suet and moss from the head of a hanged thief. The last ingredient could be omitted if the lolling dome had gathered no moss. For a hernia one physician prescribed a poultice of iron filings plus a powdered magnet swallowed by the patient. At the internal magnet's closest approach to the external iron, the hernia would be drawn back into the body.

In the 1770s a professor of astronomy called Max Hell (I blame the parents) became famous for curing all ills with magnetised metal plates applied to the body. Numerous quacks followed suit, though one revealed doubts about the methodology when he warned fellow practitioners never to magnetise in front of inquisitive persons.

The most famous magnetiser of the eighteenth century was Franz Anton Mesmer who postulated a magnetic 'fluidom' that upset the body's nervous system and could only be cured by magnetism. He claimed he could magnetise anyone with just a wave of his hand. What we call a hypnotic trance he labelled 'animal magnetism'. It was his magnetic personality and showmanship that attracted ladies to his Paris salon. The clients sat around a huge tub drinking magnetised water and caressing themselves with magnetic rods. They were then linked with cords, held hands and pressed against each others' knees to ensure that the fluence circulated.

Mesmer drifted among them dressed in a lilac robe and waving a wand like the Merlin of magnetism. Handsome youths gently massaged the ladies in their most magnetically sensitive areas until 'their cheeks began to glow' and 'their imagination inflamed'. Many swooned in 'magnetic ecstasy'.

Mesmer craved scientific recognition, but a committee of scientists including Lavoisier and Benjamin Franklin diagnosed contagious hysteria and debunked the magnetic magic. Mesmer fled in a flap of lilac.

One of his imitators used 'magnetic somnabulism' and strung his patients on a magnetised tree like Christmas decorations. In the United States Dr Elisha Perkins patented 'magnetic tractors' (magnets that draw out). When trailed over the patient's skin they extracted pain and even corrected deformities. His 'cures' were so successful that he was struck off the medical register. To prove his powers Perkins rushed to save the citizens of New York during an outbreak of yellow fever, promptly caught the fever and died.

His son Ben decamped to London and sold the magnetic tractors at five guineas each. He assured customers that 'rheumatism fled at their approach'. A sceptical doctor showed that wooden battens painted to look like metal were just as effective as the tractors. His exposé was entitled *The Imagination as a Cause and Cure of Disorder, Exemplified by the Fictitious Tractors*. It was, however, indisputable that magnetism cured poverty: Ben Perkins returned to Pennsylvania with £10,000 in his pocket and Mesmer left Paris with 340,000 francs.

They would have been astonished to learn that today electromagnets have enabled us to see the finest details of the body's interior and distinguish cancerous tissues from healthy ones. We call it Magnetic Resonance Imagery (MRI). What a money-spinner. Why didn't they think of that? Well, they would have needed magnets that could generate magnetic fields 100,000 times greater than the Earth's. Ray Damadian, the inventor of the MRI scanner, didn't know whether the human body could withstand these forces so the brave fellow became the first person to have a whole-body scan.

Such scanners also require an enormous amount of electricity, and it was the discovery of electricity that caused 'medical' magnetism to lose its appeal. By the 1730s researchers realised that they could electrify the human body, so long as it was insulated from the ground. A contemporary French etching depicts a youth suspended by ropes with his feet touching a generator of electricity – probably a rotating ball of sulphur – and sending out a spark from his fingertip to a rod held by the experimenter. The boy's face bears a pained expression.

Louis XV was intrigued and was said to have used an enormous battery to send a current through a chain of monks holding hands to make a circuit. The effect was 'prodigious'. The surprised friars proved to be great conductors of electricity and, despite their secluded life, they still knew how to dance.

The supposed therapeutic value of electricity was known to the ancients. The court physician to the Roman emperor Claudius recommended placing an electric torpedo fish on the head to cure a headache. This was inadvisable as it might have been mistaken for a very silly hat and could have delivered a charge of 220 volts that might have done away with the need to wear a hat ever again.

Eighteenth-century Britain became infatuated with electricity. A Member of Parliament with scientific leanings linked up twenty-seven volunteers to a torpedo ray and gave them a jolt they wouldn't forget. The result was 'absolutely electrical' and the sparks were 'vivid and repeated'. Quacks were in electric heaven. A Scottish physician promoted the virtues of his 'electric bath'– usually a deadly combination. Some therapies came in five colours of electrical fluids. Even the preacher John Wesley had an 'electrical machine'.

The 'Emperor of Quacks', according to the comic revue at the Haymarket Theatre, was James Graham who opened his London Temple of Health in 1780. It offered an entrancing melange of magnetism, mud baths and extraordinary thrills by, in the words of the poet Southey, 'tampering with electricity in a manner too infamous to be reported'. There was a seductively clad 'Goddess of Youth and Health' worthy of close inspection. She was certainly eyed by William Hamilton who promoted her to Lady Hamilton, later the apple of Nelson's remaining eye.

The other *pièce de* least *résistance* was the Celestial Bed. It promised 'heavenly joys' and 'superior ecstasy' as well as 'immediate conception'. The bed was wired up to give an electrical buzz to the proceedings and insulated to conserve an 'abundance of electrical fire'. Every movement of the mattress set off a symphony of organ pipes with the tempo increasing in the final movement.

The bed alone was reputed to have cost an eye-watering £10,000. Clients would have had to wear out a prodigious number of mattresses at fifty pounds a go for it just to break even. After being *the* risqué venue in London, the lavish temple went bust. Graham began self-experimenting with 'earth bathing' therapy by burying himself up to the neck in soil. He found religion but lost his reason, starving himself to increase his life span. Having promised his patrons at the Temple that they would live to be a hundred, he didn't make it to fifty.

Electrical quacks persisted well into the twentieth century. In the 1920s a rich American called Gaylord Wilshire manufactured I-on-aco, a potion of 'medical magnetism' charged from an electric-light socket. Twenty years later another American, William Reich, began dispensing an electromagnetic force called 'Orgone Energy' which came in a charming shade of blue. It was supposed to charge the body's life force and increase the sex drive. The bedridden were sold an 'energy accumulator' that produced orgone. It bore a startling resemblance to a blanket.

By this time one electrical treatment *had* become mainstream. It came not with a tingle but with a jolt. Its original name was 'Electric Shock Treatment' but that sounded rather frightening so it was changed to 'Electroconvulsive Therapy', which still sounds pretty scary to me. No doubt that's why it's always referred to as ECT.

It involves placing electrodes on the head to deliver a brief electric charge sufficient to precipitate a convulsion and render the patient unconscious. It usually causes some amnesia and is commonly used on patients with depression when all else fails.

It was not until 1980, after five decades of shocking patients' brains, that there was a comprehensive survey of the use and effectiveness of ECT. The response of *The Lancet* medical journal to the survey's findings was unequivocal: 'Every British psychiatrist should read this report and feel ashamed.' It revealed a litany of failings: poorly maintained machines, technicians with little or no training delivering shocks of arbitrary duration. This was particularly worrying as the degree of memory loss suffered is related to the amount of electricity passed through the brain. Despite studies indicating that using just one electrode rather than two caused less confusion and memory loss, eighty per cent of clinics always used two probes. ECT was even used as a 'cure' for homosexuality. Sixteen per cent of the psychiatrists interviewed admitted they would give ECT even if both the patient and relatives objected. A disconcerting sequel to this is that twenty years later a new survey of 700 patients revealed that fifty-nine per cent had been subjected to ECT without their permission.

There were 200,000 applications in Britain in 1979, the year of the first review, with each patient receiving four to eight shock treatments. Most psychiatrists agreed it was beneficial. Yet there was no understanding of *why* it might work.

Perhaps the training of psychiatrists should include at least one dose of ECT as behoves anyone who would subject others to what the first report described as a 'degrading and frightening' experience. However, if some were experimenting on their patients inadvertently, others built their reputation consciously doing so.

In the 1840s Guillaume Duchenne studied the physiology of human expressions by inducing the facial muscles to contract in ways that mirrored various emotions. His technique was to persuade a simple-minded old man to adopt a calm expression as electrodes were attached to his face.

By shocking two sets of muscles simultaneously he produced 'a strikingly truthful picture of a face stupefied with terror . . . a dreadful mixture of horror and fear', of someone awaiting 'inevitable torture'. When, in addition, he stimulated the 'muscle of pain' just above the eyes, he augmented the look of terror with an expression of agony. Duchenne took photographs that confirm his claim to have recreated 'the face of the damned'.

He called his experiments 'living anatomy' rather than the abuse of an old man by contorting the 'mundane surface' of his 'coarse face'. He could have used his own face, but perhaps his physiognomy was too refined, whereas the old chap had 'trivial features'. 'It was,' Duchenne said, 'like working with a still irritable corpse.' Duchenne had tried to reanimate the features of a cadaver, but found it far more distasteful than working on a living person. I doubt if a dedicated self-experimenter would have such a dismissive view of someone on whom he was experimenting.

Ninety years later Dr Ewen Cameron at McGill University was attracted by ECT's ability to induce amnesia. His aim was to wipe the mind clean and then re-program it. The CIA also became interested in the possibility of brainwashing and provided generous funding. Cameron 'depatterned' fifty-three of his schizophrenic patients without their consent. The process involved doses of the psychedelic drug LSD four times a week and electric shock treatment twice a day, resulting in thirty to a hundred and fifty treatments per person. This was far in excess of anything a patient would normally get.

Some patients suffering from long-term depression were kept in an ECT-induced coma for eighty-six days while being bombarded with endlessly repeated messages. Cameron called this 'psychic driving'. The patients' own words were recorded and then played back. Most were negative thoughts such as:

'I hate everything. It makes me feel so resentful. I might as well go and do something silly' – repeated thirty times.

'I hate, I hate' – thirty-five times.

'I am so lonely' – forty-five times.

Some heard the same messages a quarter of a million times. During this aural assault they became progressively more disturbed and they shook for some time after the session had ended.

By the end of the treatment the patients' memories were completely erased and some patients were mentally impaired. Their personalities had disappeared and they were unable to speak, or eat unaided. They were victims without pasts and with uncertain futures.

Cameron's own future was assured. Later he admitted that it had been a ten-year trip down the wrong path. His patients' records were destroyed. He died in 1967 replete with academic honours.

Invisible forces can be dangerous. Innumerable people have been electrocuted, often through lack of caution. Recently a man who had appeared on television demonstrating his resistance to electricity by sticking his fingers into electric sockets was electrocuted because he didn't bother to switch off the power supply before attempting to repair a generator.

We know that the high volume of sound at raves and discos is impairing the hearing of the young, but sound waves can be much more dangerous than that. Professor Gavraud often felt nauseous at work and attributed this to low-frequency vibrations from a nearby air-conditioning plant. To study this phenomenon he built a large machine driven by compressed air to produce low-frequency 'infrasound'.

Luckily Gavraud stood well back when his assistant turned on the air for the first time. The technician collapsed immediately and died. The intensity of the vibrations probably caused spontaneous breathing to cease, although it was alleged that the assistant's internal organs had been gelatinised. We are all curious by nature, none more so than scientists.

In 1895 Wilhelm Röntgen, while studying cathode rays, noticed on an adjacent bench a luminous green glow emanating from a sheet of paper coated with a barium compound. He knew this couldn't be from cathode rays which travel only a short distance in air and in any case the cathode-ray tube was covered in thick black cardboard that prevented the rays from escaping. The luminescence must therefore come from a very different invisible ray. A few simple tests showed that a thick book, two packs of cards, even a thin sheet of metal didn't block the rays. Röntgen called this powerful new force 'X-rays'.

By chance his hand passed between the tube and the paper and cast a shadow. It was not the shadow of a hand but the eerie spectre of the *bones* of his hand. This supernatural sight so unnerved Röntgen that he broke off his research. He didn't return to the laboratory until the Christmas recess when there were no other staff around. He worked feverishly and became so preoccupied that this wife feared for his mental health. She appreciated his excitement when he showed her an X-ray photograph of a lead weight inside a sealed wooden box. Although these were far more penetrating rays than anything yet discovered, his wife volunteered to be the first person to be X-rayed. This resulted in an iconic photograph of the bones of her hand wearing a ring. Dense objects like bones and metal showed up on X-rays because they partially block the rays, but only lead stops them entirely.

By January he had published an article whose title translates as 'A new kind of ray'. These see-through rays hit the headlines. There was so much clamour that the *Pall Mall Gazette* announced it was 'sick of the Roentgen Rays'. There was public concern that they might be used to see through clothes and reveal what lay beneath. One enterprising company advertised knickers that were X-ray proof.

> The Roentgen rays,
> What is this craze?
> I'm full of daze,
> Shock and amaze,
> For nowadays,
> I hear they'll gaze
> Through cloak and gown and even stays,
> These naughty, naughty Roentgen rays

Within months of Röntgen's publication X-ray machines had been designed and built and were being used to look for fractured bones and gallstones. Before long bismuth and then barium were being swallowed to make the interior of the gut visible to X-rays. The Horlicks company made the barium meals.

It was soon apparent that the rays could burn the skin, but there was much worse to come. X-rays are a form of ionising radiation that can cause fundamental damage to the body. They can kill cells or alter them irreversibly. This can lead to sterility, cancer and damage to the DNA, with dire repercussions for future progeny. The damage is dose-related – the more exposure, the greater the risk. As early as May 1897 there were sixty-nine reports of X-ray injuries, all of them to technicians working with X-ray machines. Whether they knew it or not they were conducting deadly experiments on themselves.

Thomas Edison worked with the machines but stopped 'when I came near to losing my eyesight'. Clarence Dally, one if his assistants, fared worse. He developed ulcerated arms and incredibly was given more X-rays to 'heal' the damage. His arms had to be amputated and he died a year later at the age of thirty-nine.

It was not a pleasant way to die, as exemplified by the fate of Mr Cox who maintained the army surgeon's X-ray machine during the South African War. A friend visited Cox and described his condition: 'A more pitiable case I have never witnessed in any human being.' He spoke of the 'excruciating agony he has endured for the past six years. Words cannot depict the awful condition of the man . . . Poor Cox, whose shattered health is beyond all doubt irrecoverable, has lost his right arm and fingers from his withered left hand and was totally incapacitated.' He had also developed 'rodent ulcers and a most serious cancerous condition attacking his face, chin and jaw'. It took his wife two hours to dress his 'terribly affected face'.

The medical journals began to catalogue the radiologists and radiographers who were 'sacrificed to science'. They included Walter Cannon who had been the first to use radio-opaque chemicals as well as pearl buttons to study the internal workings of throat and stomach. Röntgen was spared, perhaps because he kept his X-ray tubes in a metal box.

Technicians were in the firing line because they wore no protective clothing and the early X-ray tubes had little or no shielding. Patients were also at risk from being given excessive doses of irradiation. In 1912 the X-ray tube was described as 'a remarkably fickle appliance and it was quite impossible to estimate the magnitude of the dose'. The duration of exposure was incredibly long to ensure a good image. In the first account of finding a foreign object (a bullet) inside a human, published in 1896, the exposure took more than two hours. An abnormally hirsute five-year-old girl was treated with X-rays for two hours a day over sixteen days. The hair on her back did indeed fall out only to be replaced by a huge necrotic ulcer destined to become malignant. Such treatment left tens of thousands of women scarred or cancerous. The last X-ray machine dedicated to depilation was not decommissioned until 1949.

Countless thousands of people were irradiated unnecessarily. In the early days photographic studios offered 'bone portraits' and even when I was a child many shoe shops had a 'pedoscope' that X-rayed your feet inside the shoes to show whether they were a good fit. Kids loved pressing their eyes against the glass viewing plate to see their skeletal toes.

In 1906 a man demanded that his fiancée was X-rayed before the wedding to confirm she was healthy inside – possibly because he had not been granted access to her interior during their courtship. She refused and he broke off the engagement. She sued for breach of promise and won generous compensation.

No sooner were X-rays available than they were used in court to support a compensation claim for an injury suffered by a burlesque actress, to prove a surgeon's negligence, a murderous husband's guilt, and to reveal that a suspicious parcel contained a terrorist's bomb.

A detective even proposed their employment to prove infidelity in divorce cases. An electrical magazine pondered how. 'We assume he uses X-rays to discover the skeletons which every closet is said to contain.'

A judge refused to accept X-ray photos as legal evidence since: 'There is no proof that such a thing is possible. It is like offering the photograph of a ghost.' In December 1896 a Denver attorney argued for three hours that they should be inadmissible. The opposing lawyer brought an X-ray machine into the courtroom and X-rayed the jurors' hands to prove the technique's bona fides. The trial judge ruled that: 'Modern science has made it possible to look beneath the tissue of the human body . . . We believe it to be our duty in this case to be the first . . . in admitting in evidence a process acknowledged as a determinate science.'

The laws that interested Marie and Pierre Curie were those of physics and chemistry. In the same year that Röntgen discovered X-rays, Pierre received his doctorate for research on magnetism and married Marie Sklodowska, a Polish student who had been starving in a Parisian garret normally reserved for artists. They formed one of the greatest partnerships in the history of science.

Henri Becquerel had just discovered a new form of radiation emanating from uranium. They decided to investigate this new ray. Marie worked unpaid and the only room she could have had been abandoned by the medics as too sordid even for dissecting cadavers. A visitor described it as 'a cross between a stable and a potato shed'.

Marie found that different compounds of uranium all emitted radiation, so clearly it had nothing to do with their chemistry: it was a property of the atom itself. She called it 'radioactivity'.

Uranium was found in an ore called pitchblende, which had been surreptitiously killing miners for centuries. When the first tonne of pitchblende arrived at Marie's laboratory she was so thrilled that she thrust her hands into the brown ore and let it sift between her fingers.

Marie had devised a way to measure the amount of radiation. To her surprise the pitchblende was far more radioactive than any amount of uranium it might contain could explain. So she searched for another highly radioactive constituent. It was a tougher task than she anticipated as its concentration in the ore was only five parts in a billion. Nonetheless she isolated it and named it 'polonium' after her native Poland. But even this didn't account for all of the radioactivity. She was so sure there must be yet another source of radiation that she called it 'radium' even before she found it.

Isolating the radium involved repeatedly boiling and cooling solutions in large vats to get progressively purer crystals. Marie often 'passed the whole day stirring a mass of ebullition with an iron rod as big as myself' until she was 'broken with fatigue'. Eight tonnes of pitchblende yielded only a tenth of a gram of radium.

She had discovered two new elements and became the first woman in France to receive a doctorate. Like Röntgen she did not patent any of her discoveries as it would have been 'contrary to the scientific spirit'.

At night the luminescent tubes of radium on the shelves 'looked like faint fairy lights'. According to Pierre the blue glow 'was sufficient to read by if the tubes were held near to the page. And of course it was a *very* small quantity.' Radium was both captivating and lethal. It gave out heat as well as light and was a million times more radioactive than uranium.

Marie suffered skin burns after carrying one of the vials

in her pocket. Pierre showed that radium was the culprit by strapping some to his arm. He developed recalcitrant ulcers. Pierre also carried out tests on mice and guinea pigs, exposing them to the radon gas that radium gave off. They all died. Surely they must have now realised that if they continued to experiment with radium, even breathing the air in the laboratory would be hazardous. But the excitement of their discoveries drove them on. Marie once said, 'Nothing in life is to be feared. It is only to be understood.'

In 1903 they shared the Nobel Prize for physics but were unable to attend the ceremony in Stockholm because of Marie's anaemia. It might have been the first symptom of radiation exposure. She also suffered a miscarriage.

When Pierre was heating a tube of radium to a high temperature it exploded, scattering its dangerous contents everywhere. He subsequently developed problems with his eyes and such crippling pains in his legs that he was unable to work. In 1906 he fell under a dray and one of its wheels crushed his head. He died instantly.

Marie was distraught, but eventually found solace in her research: 'I could not live without my laboratory.' She was invited to take Pierre's place as Professor of Radio-physics at the Sorbonne. It was unprecedented for a woman to hold an academic chair so for the first couple of years her title was *Chargée de cours* before she was elevated to 'professor'. In 1911 she became a Nobel laureate again, this time for chemistry, and became the first person to win two Nobel Prizes in different disciplines. Even this didn't get her elected to the French Academy – she was, after all, a woman.

During the First World War Marie was instrumental in providing mobile X-ray units for the troops. She even drove one herself and trained the operators. By 1918 there was an X-ray unit in every field hospital and it was estimated they had treated a million wounded soldiers.

In the 1920s Marie's health deteriorated. She developed painful radiation burns on her hands and had several operations to save her sight. Eventually she succumbed to severe anaemia resulting from radiation damage to her bone marrow where blood cells are manufactured. Her lasting memorial is the 'curie', the primary unit of radioactivity. She would have been delighted that radiotherapy using radium became a major weapon in the fight against cancer, but perhaps dismayed had she known that polonium would be used to detonate the atom bomb and to assassinate a Russian dissident in London in 2006.

In 1995 it was decreed that the Curies should be disinterred and reburied under the grand dome of the Pantheon in Paris. Both President Mitterrand of France and Poland's Lech Walesa attended the ceremony. An enterprising scientist took the opportunity to scan the bodies with a Geiger counter. Both were radioactive, Pierre astonishingly so. His tragic traffic accident had spared him from a terrible, lingering death.

Marie's daughter Irene followed in her mother's footsteps. She became a scientist, shared a Nobel Prize with her husband

and died of leukaemia, probably caused by working with radiation.

Over a period of six years George Stover, an American radiologist, deliberately tested the effects of radium on his own body. As a consequence he later needed several amputations and over a hundred skin grafts. Shortly before his premature death he said, 'A few dead or crippled scientists do not weigh much against a useful fact.'

Such horror stories had little impact on the general public compared with the news that radium was being used to cure cancer. It became known as the 'ray of life'. It became the magic ingredient in dozens of products: chocolate, toothpaste, radium salves 'to make the skin tingle', radioactive clothes to keep you warm, hearing aids with the little-known element 'hearium', Ra-Ba-Sa bath salts and hair restorers, even though radiation makes your hair fall out. There were radioactive lotions, douches and suppositories eager for insertion. No cranny of the body was safe. Raditone tonic tablets with 'gland extracts' 'increased sexual vigour' and there was a radioactive contraceptive gel just in case they did.

Most of the products were probably harmless. Since radium was far more valuable than gold, they probably didn't contain any. However, radium gives off radioactive radon gas for a thousand years and even a few specks produce a significant amount. Sparklets promoted carbon dioxide bulbs spiked with radon for making soda water in their siphons. There were various 'emanator' jugs and jars that sold like hot cakes of uranium. Water left to stand overnight became charged with radon. 'All next day the family is provided with two gallons of healthful radioactive water . . . Nature's way to health.'

William Bailey, a serial fraudster peddling 'high-priced hokum', hit the jackpot in the 1920s with radium-rich Radithor water, selling 100,000 bottles a year. It was 'the fullest achievement in internal radioactive treatment'. He had never said a truer word.

One of his best customers was a rich socialite and sportsman called Eben Byers. He was heading for fifty and in need of reinvigoration. Radithor was the answer. He was so pleased with it that he sent crates of the stuff to his friends. Then he began to have constant headaches, his teeth fell out and his bones became brittle.

When we think of radioactivity we imagine gamma rays that pass through the body like a bullet damaging every internal organ, and from which our only defence is a lead wall. But there are also beta rays that penetrate only a few centimetres into the body, and alpha rays, which are relatively slow (a mere 15 million metres per second) and are blocked by a sheet of paper or our skin. But alpha radiation is lethal if ingested. Radium has a proclivity for bones where it accumulates and the alpha rays bombard the surrounding bone to a honeycomb. They never run out of ammunition – the victim runs out of bone.

A lawyer intent on prosecuting Bailey came to interview

Byers and described his predicament: 'A more gruesome experience would be hard to imagine. Byers' condition beggars description . . . he could hardly speak . . . his whole upper jaw and most of his lower jaw have been removed. All the remaining bone tissue of his body was slowly disintegrating, and holes were actually forming in his skull.' Byers died in agony.

Bailey was successfully prosecuted, but only for making fraudulent claims. He was soon back in business selling Health Springs, 'rich in vital rays', and Bioray paperweights, 'far richer in the invisible gamma rays than the sun'. There was no law to prevent him selling what he called 'perpetual sunshine'. He died of bladder cancer brought on by an excess of it.

Radium's most marketable property was its luminescence. Undark manufactured paint kits with which you could illuminate your door's bell push, or a gunsight, fish bait, even 'a woman's felt slippers'. The company's founders envisaged future homes being illuminated by radium-painted ceilings, but the demand was for important things that needed to glow in the dark such as dolls' eyes, rouge and cocktails.

With the coming of the First World War there were innumerable dials that had to be read in dim light. It was estimated that ninety-five per cent of radium production in the United States was devoted to the production of luminous paint, at the expense of medical usage. Some doctors even sold some of their allocation. By 1917 every American serviceman had a watch with a luminous dial.

The paint was radium mixed with tiny crystals and sweet-tasting gum arabic. Painting the numerals and hands on watches was fine work. The brushes had only three or four bristles. To 'point' the bristles or to remove excess paint the artists wiped the brush between their lips. Painting 250 dials a day meant they swallowed a lot of radioactive material. One factory employed 800 female dial painters over a seven-year period. By the 1920s there were 120 such factories in the United States.

A dentist was the first to notice that several of the factory girls had jaw problems, and in 1925 the resident physicist at one factory died and an autopsy revealed severe necrosis of his lungs and liver. His bones were so radioactive that when left in the dark on a sensitive plate they photographed themselves. Shortly afterwards one of the dial painters was admitted to hospital. Even her breath was radioactive. She died of severe anaemia because her bone marrow had been destroyed. Not a single woman who had worked at the factory for a year or more had a healthy blood count. Some were terrified to find they were luminous in the dark.

The management dismissed suggestions of 'so-called radium poisoning . . . Nothing approaching such symptoms have ever been found.' Yet they had just suppressed a highly critical health report and their dead employees were being buried in lead-lined caskets. The company president wrote a sanctimonious letter to the local Commissioner of Health pleading that the deaths were because it was company policy to hire those whom others would consider unemployable. 'What was considered an act of kindness on our part has since been turned against us.' Sick workers took the company to court and won compensation, but the women with rotting jaws and honey-combed bones continued to die for years afterwards.

The clocks and watches they painted lived on. It was esti-mated that during the years they spent on wrists and bedside tables they dispensed more radioactivity than all the radium-processing sites and nuclear facilities in the entire USA.

Fortunately the fad for radioactivity died a natural death, which is more than can be said for its victims. Yet even as recently as the 1980s Soviet doctors are said to have prescribed 2,500 radon baths daily, and the Radium Palace Hotel near Karlsbad still welcomes visitors to its 300 bedrooms and its naturally radioactive springs. Fortunately Buxton Spa in England no longer imports radioactive water so that it can't boast as it once did that its waters were fifty times more radioactive than even the 'hottest' places abroad.

Radiation lingers. Marie Curie's notebooks are still radioactive. In 2008 the University of Manchester in England grew suspicious about the cause of fatal cancers in staff who had recently worked in the building where Ernest Rutherford had first described alpha radiation almost a century earlier. Rooms thought to be contaminated were sealed off and all staff who had worked there over the previous twenty years were advised to have a health check.

Another person having his health monitored was Eric Voice, a nuclear chemist. In the mid-1990s he and eleven colleagues were injected with a short-lived isotope of radioactive plutonium to determine where and for how long it remained in the body. This was to inform estimates of the risks involved from an accidental exposure to larger quantities. In 2000, though now retired, Voice inhaled plutonium to test how much was absorbed by his lungs and how quickly it travelled around the body. He boasted that he was now the world's most radioactive man. An armoured van called regularly to collect his waste, although he showed no adverse effects of his exposure.

Already in his seventies, Dr Voice was phlegmatic about his fate: 'I don't think I need to worry too much about what could happen in the rest of my life.' He was one of those rare people who did not want to live as long as Methuselah. He died of motor neurone disease in 2004, aged eighty, and was buried in a lead-lined coffin.

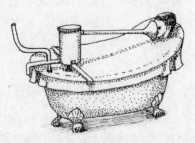

Health spa heaven. A radium bath with radon inhaler.

Found to be Wanting

'People have long argued about the location of Hell. We have discovered it' — Louis Antoine de Bougainville on a scurvy-stricken ship

We can readily appreciate the dangers that might ensue from viruses, drugs or radiation, but terrible conditions also come from something we lack.

For centuries a sailor's life was not a happy one. He was probably pressed into service against his will and then confined in a seagoing prison with the option of drowning. But more mariners died of scurvy than in all the storms, wrecks, battles and infectious epidemics combined. During the Seven Years War, for every man who was killed in action eighty-nine died from disease, mostly scurvy. In the age of sail over two million sailors are thought to have died from this terrible affliction. A lucky ship was one with enough fit crew to make its way home.

In 1740 Commander George Anson was ordered to 'annoy and distress' the Spanish in the Pacific. The venture did not begin well. There were insufficient men to crew his ships so the Admiralty decanted pensioners from the Chelsea Hospital to make up the shortfall. Anson received a tottering flock of 'crazy and infirm . . . decrepit and miserable objects . . . fitter for the infirmary than any military duty'. Even the most able-bodied would be fortunate to survive this fated mission. Only three men were lost in action, yet of nearly 2,000 that set sail less than 200 returned. Almost all had succumbed to scurvy.

Every ship left corpses bobbing in its wake. It was worse for the ships from Catholic Spain and France because their dead had to be buried in consecrated ground so cadavers were stowed among the ballast, to slop around in a lake of corruption. Some crewmen were asphyxiated by the foul vapours seeping from below.

Scurvy was a terrible condition. Capillaries leaked beneath the skin, the gums turned to swollen sponges exuding putrid blood and liberating the teeth. Even long-healed fractures of bones fell apart. According to Anson, 'some lost their senses . . . and some rotted away'.

The ordeal of Anson's voyage shocked the government into searching for a cure for scurvy. But the solution had been known for over a century. 'Lemon water juice' was an effective preventative used on voyages of the East India Company. The Company had even planted citrus groves at staging posts on major shipping routes. However, within thirty years complacent captains baulked at buying expensive fruit for a condition that was now seen only occasionally. They were encouraged by an 'expert' who claimed that fruits from exotic places were the *cause* of fevers such as scurvy. The cure was forgotten and replaced by quack remedies. The dread disease returned.

In 1753 a young Scots naval surgeon called James Lind challenged the theories on the cure of scurvy 'invented according to the whim of each author and the philosophy then in fashion'. Instead of theorising he set up a test on board HMS *Salisbury*. It was one of the first clinical trials under controlled conditions. He chose two dozen men showing similar symptoms of scurvy. Half of them were split into groups of two and each pair received a different 'remedy' including such horrors as elixir vitriol (a refreshing mix of alcohol and sulphuric acid), food drenched in vinegar with a nip of neat vinegar to follow, or a pint of seawater a day. The other twelve men received no supplements and were the untreated 'control' group. Lind ensured that they all ate the same food and lived separately from the rest of the crew so that he could keep an eye on them. Although the sample size was only two, Lind concluded that, of the treatments tested, 'oranges and lemons were the most effectual remedies for this distemper of the sea'. Lind is now called 'the father of nautical medicine' but although he wrote a comprehensive report of his findings no one at the time took the slightest notice.

In 1768 James Cook set out on his first expedition to the Pacific and the Southern Ocean. Part of his remit was to critically test different treatments for scurvy. He enforced the use of dietary supplements although the crew disliked 'novel' foodstuffs. They happily munched their way through ship's biscuits riddled with weevils and salted meat infested with maggots, yet some had to be flogged for refusing to eat *fresh* meat picked up at Madeira. They refused sauerkraut so Cook craftily had it on the officers' dining table every day, knowing that when they saw 'their superiors set a value on it, it becomes the finest stuff in the world'. They also grumbled about the bitter taste of lemon juice, but it was all worthwhile. Thanks to his strict dietary regime and gathering fresh vegetables (including scurvy grass) at every landfall, Cook returned home after spending three years at sea without losing a single hand to scurvy.

The Admiralty were delighted with his explorations but unconvinced by his dietary findings. Because he had used several supposed remedies such as carrot marmalade and wort of malt, it was unclear which one had controlled the disease. The Admiralty didn't take their advice from captains who confronted scurvy on every voyage; instead they consulted the learned gentlemen of the Royal Society, who had probably never seen a case of it. They prescribed useless nostrums based on their erroneous ideas of the nature of disease.

The Admiralty made clear that there had been trials 'on

board several different ships which made voyages around the globe . . . the surgeons of which all agree in saying that the "rob" [syrup] of lemons and oranges were of no service in the prevention, or cure of that disease'. The problem was that Lind and others had recommended the use of 'rob'. It was a concentrate of orange or lemon made by evaporating their juice over heat. This halved the potency of the juice which rapidly declined even further when stored. Only *fresh* fruit did the trick, for the active ingredient was denatured by heating or drying.

Dismissing the only effective treatment allowed scurvy in the British navy to reach epidemic proportions. Other navies were similarly afflicted. In 1770 a combined Spanish and French force was to launch an invasion of England. The Spanish ships were delayed for just seven weeks, by which time two-thirds of the French seamen were incapacitated by scurvy. The only invaders that reached England were the vast numbers of bodies tossed overboard to wash up on the south coast.

Ten years later the entire crew of a British frigate was laid low with scurvy. Their lives were saved by a passing American ship whose master gave them fresh meat and vegetables. The frigate's twenty-two-year-old captain was Horatio Nelson. Had he died, what would Britain have done without her greatest commander?

That same year Gilbert Blane, personal physician to Admiral Sir George Rodney, expressed dismay at the health of British sailors. He was convinced that 'more can be done towards the preservation of the health and lives of seamen than is commonly imagined'. After reading both Lind and Cook's reports he proposed that orange juice be distributed to every man in Rodney's fleet. Their death rate from scurvy plummeted from twenty-five per cent a year to five per cent. Later, when he was appointed to the Admiralty's Sick and Hurt Board, Blane used his influence to improve the hospital services ashore and to ensure that citrus juice reached the entire navy. Soon Britain had the healthiest sailors in the world, as Napoleon was about to find out.

During the Seven Years War and the American War of Independence the Royal Navy Hospital at Haslar routinely treated 350–1,000 scurvy patients per day, rising to 2,400 on a single day in 1780, from ships docking at Portsmouth. By 1815 the hospital had seen only two cases of scurvy in four years. In the forty years that had elapsed since Lind's experiment hundreds of thousands of mariners had died unnecessarily.

Not all the casualties were sailors. A young physician called William Stark became acquainted with Benjamin Franklin who was living in London at the time. Franklin mentioned that he had once stayed healthy while taking nothing but bread and water for two weeks. Stark decided to test which simple foods were most beneficial and which were not. With 'impudent zeal' he launched into a nine-month period of restricted diets; ten weeks on bread and water were followed by a stint on flour, bread and oil, then a comparative study of the effects of lean versus fatty meat. He had intended to sample fruit and green vegetables next, but tragically chose cheese and honey instead.

Stark rapidly sank into the clutches of scurvy. A distinguished physician familiar with scurvy merely advised him to reduce his salt intake. Stark died soon afterwards at the age of twenty-nine.

Although it was accepted that something in citrus fruits cured scurvy, it hadn't occurred to most of the pioneers that it was the *lack* of that something that caused the condition. Sailors (and the poor) were condemned by their diet. If a person is given only sufficient protein, fats, carbohydrates and water, they will surely die. Other compounds that we now call vitamins are vital for our health.

Ascorbic acid, vitamin C, is the one required to avoid scurvy. It is abundant not just in citrus fruits but also in rose hips, blackcurrants, papayas, kiwi fruit, broccoli, Brussels sprouts, parsley and watercress, but not in all fruits and vegetables. There is little or none in meat except for liver and kidneys. Ascorbic acid is essential for the formation of collagen, the most abundant protein in the body. Collagen is the fibrous basis for the 'cement' that binds cells together. Without it the body literally comes unglued.

Breast milk is also rich in vitamin C, 'stolen' from the mother. In Victorian times, when many of the middle class switched from breast to bottle feeding, their babies developed early symptoms of scurvy as the milk they used contained no vitamin C.

The curse of scurvy was virtually confined to humans because almost all other animals, including most apes, can manufacture vitamin C in their bodies and need no external supply. In 1907 researchers showed that if guinea pigs were fed exclusively on grain (lacking vitamin C) they developed scurvy-like symptoms, which then vanished if they were given the vitamin. The researchers were lucky because guinea pigs are one of the very few animals that, like us, need an external source of vitamin C.

The action of vitamins was largely determined by dietary experiments on animals and human volunteers – although they didn't always know they were volunteering. Scurvy had been common in institutions such as prisons and orphanages, thanks to an inadequate diet. In the better orphanages the children were given orange juice. In 1913 two paediatricians, Alfred Hess and Mildred Fish, withheld orange juice from orphaned babies until they developed the localised bleeding lesions characteristic of scurvy. The idea was to use the infants to try out an invasive diagnostic test that involved puncturing the abdomen. They then repeated the experiment to see if they could induce the disease for a second time. They also used a similar protocol to study rickets, a bone disease caused by a deficiency of vitamin D.

The researchers felt no qualms about these experiments. It was stated that the orphans had made 'a large return to the community for the care devoted to them'. There was general satisfaction that they were repaying their debt to society. Experimentation on infants (especially orphans and 'mental defectives') was commonplace well into the first half of the twentieth century. 'Infant volunteers', as they were called, came with the convenience that only the permission of the orphanage director was required to 'use' them. As babies had not lived long enough to have developed immunity to most diseases they were ideal for testing new vaccines. The procedure was to inject the experimental vaccine and then expose the infant to the target disease to see if immunity had been induced. These human guinea piglets were 'challenged' by the *live* agents of whooping cough, measles, hepatitis, smallpox and herpes. Dr John Kolmer tested a live polio vaccine on himself and his two boys as a reassuring prelude to injecting three hundred children. After nine deaths the vaccine was withdrawn. In 1930 in Germany seventy-six infants died while undergoing a vaccine trial for tuberculosis.

John Crandon, a young surgical resident at Boston City Hospital, was not the sort of man to experiment on children. In 1939 he became interested in why poorly nourished people took longer to heal. Was there a clue in the observation that old wound scars broke open in scurvy sufferers? He designed an experiment in which the subject would be 'prepared' on a diet deficient in vitamin C so that the healing time of experimentally inflicted wounds could then be determined. Crandon was adamant that he would set a good example by being one of the guinea pigs. Two other volunteers for the project were dropped when they were caught quaffing illicit orange juice.

Three months into the experiment a deep wound was cut into one side of Crandon's back, with another cut into the other side later on. Crandon was still on what was virtually a cheese, crackers and coffee diet. He had lost fourteen kilos and was close to exhaustion. The laboratory director voiced fears that he would kill himself. And he nearly did.

While exercising in the gym in an attempt to keep fit, Crandon's pulse soared and, with the conviction that he was dying, he passed out. It was known for scurvy patients to die suddenly from internal bleeding. His scurvy was sufficiently advanced for the scar from an appendectomy operation to reopen. He had had the operation when he was eight years old. The experimental wounds showed no sign of mending until he was given injections of vitamin C. He had shown that the low healing capabilities of the poor that invited infections could be rectified by more vitamin C in their diet.

Crandon's mother Margery was a renowned psychic who was outed as a fraud by Houdini. Her son, on the other hand, was the genuine article. A courageous self-experimenter.

Crandon was not the only researcher to deliberately induce vitamin deficiency in his own body. In the same hospital twenty-two years later a young haematologist almost died while doing so. Victor Herbert came from a family of achievers and his contribution was that he developed the first test for folic acid (one of a complex of B vitamins) in human blood. It is found in a variety of foods and especially in leafy green vegetables, but like vitamin C it is destroyed by cooking. It was assumed that since bacteria in the gut were thought to synthesise it, folic acid deficiency was not possible except in those with serious digestive problems and in alcoholics, as alcohol degrades this vitamin.

One of Herbert's patients had both scurvy and megaloblastic anaemia, a serious condition in which the red blood cells are abnormal. The patient was a man of habit. Like many senior citizens who have to survive on limited means he neglected his nutrition. He shunned vegetables and fruit, indeed almost all the foods that would be included in a healthy diet. His sole source of protein was from cheap, chronically steamed hamburgers. It was a case of slow hamburger homicide of the most effective kind.

Folic acid was implicated in the production of red blood cells and Herbert surmised that steaming the burgers had destroyed all traces of folic acid – hence the patient's anaemia. To test whether this was possible he needed to show experimentally that folic acid deficiency could produce megaloblastic anaemia. In the time-honoured way he invited one of his juniors to try the experiment. Though flattered, the junior wisely declined. Herbert rationalised doing the experiment on himself: 'If anything went wrong with someone else, I would not want it on my conscience.'

The first task was to prepare a folic-acid-free diet. This involved boiling everything until all traces of folic acid, taste and texture were extinguished. Even lashings of sauce couldn't prevent finely diced, thrice-boiled chicken tasting like nothing at all.

Herbert stuck to the diet for seven months. The only 'extras' were tablets to replace the *other* vitamins that had been destroyed in the boiling. The sheer monotony of it blunted his appetite to the extent that he lost twelve kilograms. He became emaciated, befuddled and disorientated, but he boasted that while those around him sniffled and sneezed, he seemed immune to the common cold.

His blood cells were repeatedly checked and samples of bone marrow (where blood cells are made) were taken by pushing a large needle into his breastbone. It is a painful procedure and the first time Herbert was terrified that the assistant would press too hard and the needle would be thrust into his heart. Several people had died from complications arising from just such a mishap. He had only eight more ordeals by needle to go.

So far there was no sign of anaemia, but there was an unexpected and more disturbing development. One morning Herbert awoke to find that he couldn't get out of bed. Had the lack of folic acid permanently paralysed his legs? By chance he had recently read a scientific article about paralysis induced by a shortage of potassium, for among its many other virtues potassium regulates the functioning of nerves. Had boiling his food depleted the potassium content too? It had. His potassium levels were so low that he could have dropped dead at any moment.

With the help of a potassium tonic the paralysis disappeared, but there was another trial to come. In order to secure a piece of tissue from the gut lining to examine under the microscope, a long tube was passed down Herbert's throat into the small intestine. Blades on the end of the tube were supposed to excise a tiny sample. When the surgeon began to haul the tube back up, Herbert screamed. It felt as if his guts were being dragged out. And they were. Somehow the blades had clamped onto the gut wall and wouldn't let go. For a moment it looked as if the sample might be the entire intestine.

At last, after seven months, Herbert developed megaloblastic anaemia. His hypotheses had been correct and he could abandon his diet and tuck into a nice plate of folic acid. Like many other self-experimenters he did not boast about his risky research. In his publications the guinea pig is an anonymous male physician.

Folic acid deficiency, having been overlooked for decades, is now recognised as a worldwide problem. Thanks to the detailed monitoring of Herbert's condition a simple blood test can now determine how advanced anaemia is in a patient.

Folic acid is essential for the formation of both red blood cells and genetic material. It therefore plays a major role in the development of a foetus, which drains the mother's reserves. Today pregnant women are given folic acid supplements to replenish their supply.

We rightly think of vitamins as being a 'good thing', but you can have too much of a good thing. Vitamin A protects us from infection, but the body needs only 0.003 grams a day – a mere speck is sufficient. We cannot excrete any excess so it is stored within the body and it's poisonous.

Carrots are a rich source of vitamin A and I recall an inquest on a man who was found dead with a strange rash over his body. His hair had fallen out and his skin was bright orange. It transpired that he had lived almost exclusively on carrots. He consumed carrot soup, carrot casserole and carrot cake. His daily tipple was, you've guessed it, carrot juice. He had been killed by an excess of vitamin A.

The coroner asked what the victim did for a living. The reply was, 'He was a research biologist, sir.'

'Ah . . .' said the coroner as if that explained everything, and perhaps it did.

As a schoolboy I was morbidly fascinated by a textbook's photographs of people with unpleasant conditions such as parasitic elephantiasis, Klinefelter syndrome (males with an extra X chromosome) and a vitamin-deficiency disease called pellagra. The boy in the photograph was about the same age as I was. Whereas I fretted over the occasional zit, all the skin on his chest, hands and face was corrugated by scaly crusts.

Pellagra is even worse than it looks. It begins as dermatitis, followed by bloody diarrhoea due to ulcers in the bowel. Later on dementia sets in, followed by death. The boy in the photograph was doomed, yet all he needed was a diet of milk and cheese and lightly cooked fish, food rich in niacin, the vitamin he lacked. But in his time no one knew about vitamins. Even after all these years, I still remember his large, uncomprehending eyes.

An outbreak of pellagra in an Alabama lunatic asylum in 1906 made the mad even more insane and killed 101 inmates. It had always been common in poor rural districts and now an epidemic swept through the southern states of the USA. It spread to the cities because wholegrain meal was being replaced with finely milled grain from which the niacin had been inadvertently removed. Food processing often inactivates vitamins.

In 1914 Dr Joseph Goldberger was asked to head the Pellagra Commission that was attempting to control the disease. It was assumed to be infectious, but its behaviour was peculiar. In England two brothers had recently died of pellagra. One was dead within four months of the first symptoms appearing, although he had spent most of his life with his elder brother who had been chronically ill with pellagra for at least six years. Also although inmates of prisons, asylums and orphanages were often ravaged by the disease, their warders and carers were not.

This didn't sound like an infectious disease to Goldberger. Researchers had failed to infect monkeys with pellagra, but what about humans? Goldberger, his wife and dozens of other volunteers tried every means to give themselves pellagra. Instead of seminars they attended 'filth parties' where Mrs Goldberger injected herself with blood from a woman dying of pellagra. The men took mucus from the noses of infected patients and smeared it into their own noses and mouths. Babies have a tendency to put everything in their mouths — so have self-experimenters. The transfers were tried over and over again until they could take no more. Not one of them developed pellagra.

Goldberger ran a clinical trial at a prison. A dozen healthy men including a murderer volunteered. They were given a nutritious but limited diet. After six months, six of the remaining eleven had developed pellagra. The twelfth man, an impetuous fellow, climbed the prison wall and escaped to become a wanted man. In his excitement, he had forgotten that his reward for completing the trial was a free pardon. At the end of the trial all those who contracted the disease refused free medical aftercare and 'went off like scared rabbits'.

These volunteers had been well rewarded, but convicts were cruelly exploited elsewhere. According to one doctor they were 'much cheaper than chimpanzees'. Indeed, they could be profitable. In a US prison the doctors received $300,000 from a pharmaceutical firm for running a trial. They augmented this payment by bleeding the felons almost every week and selling the blood.

Goldberger went on to cure an entire ward of pellagrous

children by simply diversifying their diet. Pellagra continued unabated in the adjacent, untreated ward.

Later his team showed that brewer's yeast (which is rich in niacin) cured pellagra. Even a few patients who had reached the dementia stage recovered their wits. As with many of the vitamin-deficiency diseases, the cure preceded the identification of the causal agent.

Goldberger's findings saved hundreds of thousands of lives and lifted a pall of misery. He was nominated five times for the Nobel Prize, but the invitation never came.

Since we discovered vitamins we have taken them and other dietary supplements in ever-increasing amounts. Instead of relying on a varied diet, we are beginning to rattle with pills. Britain alone has twenty million 'vitamin regulars'. I read of a young woman who spends £100 a week on dietary supplements. She confessed: 'I take so many vitamins each morning that I never fancy breakfast.'

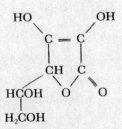

Chemical structure of ascorbic acid, Vitamin C.

Something in the Blood

'Blood is a most peculiar fluid, my friend' — Goethe

Blood is the body's lifeline. It transports oxygen and the nutrients from our food to every needy organ and tissue. It also dispatches hormones to their place of action and is the supply line for our main defenders against disease. A pinprick of blood contains 7,000 white blood cells eager to engulf alien invaders.

Even the earliest barber-surgeons bestowed great importance upon blood, but instead of trying to retain it within the body they were determined to facilitate its escape. For this we should perhaps blame Galen, a second-century AD Roman doctor who learned his trade mending gladiators and became the Emperor's physician. He was also the know-all of medicine and his influence lasted for well over a thousand years. Galen decided that all headaches, fevers and apoplexy were caused by a build-up of blood. Draining was the obvious remedy. From then on the bleeding bowl was always to hand and even into the nineteenth century blood-letting was the prime treatment for everything from syphilis to madness. The other physician's favourite was purging, both oral and anal, so the poor patient got either a knife into his vein or a funnel up his bum – or both.

The simplest bleeding technique was to tie a bandage around the arm, causing the vein to swell, then to lance it, allowing blood to spurt into a bowl. It was called 'breathing the vein'. Leeches were also the physician's little helpers. If placed on the skin they bite in and suck away until swollen to four times their original size. When replete they fall off, but if detached prematurely they leave their teeth behind to fester. Leeches inject an anti-coagulant to aid the flow of blood and it continues to encourage bleeding for hours after-wards. Leeching became so popular that thousands of ponds were stripped of their slimy suckers. In 1837 France imported 33 million leeches to meet demand.

'Cupping' also became popular. It involved placing a heated glass jar over a cut made in the skin. As the air inside cooled, it contracted, forming a vacuum that sucked blood from the wound. Cupping was commonly used to treat high blood pressure until the 1950s.

Surprisingly, many patients looked forward to regular bloodlettings. Beneath an eighteenth-century drawing of a fine lady being bled by a surgeon the caption reveals her feelings: 'Courage, Sir. I'll be brave . . . Puncture with confidence. Make a good opening. Ahh, the gush of blood surprises you ... Oh Gods, the gentle hand, the agreeable puncture ... blood drawn makes me feel much better ... I sense my vigour return anew.'

The trouble was that 'heroic' bleeding became the norm. An early-eighteenth-century novel by Alaine le Sage describes a surgeon treating a canon with gout. After extracting 'six good porringers of blood', he instructs his assistant: 'You will take as much more three hours hence, and tomorrow you will repeat the operation. It is a mere vulgar error that blood is any use to the system; the faster you draw it off the better ...' This regime 'reduced the old Canon to death's door in less than two days'.

It was fiction, but no exaggeration. Benjamin Rush, 'the founding father of American medicine', believed that 'hypertension of the arteries' was the key to disease and the cure was copious bleeding. The nineteenth-century French surgeon Broussais takes the prize for being the greatest bleeder of all time. For efficient draining he used battalions of fifty leeches attached all over the body. Historians claim that he shed more blood than all the wars during his lifetime. It's estimated that he might have siphoned off 20–30 million litres (35–50 million pints) during his career. His protégé Jean Bouillard also drained his patients dry, both bloodwise and financially. By repeated bloodletting, taking up to three litres, he ensured that the suffering of patients was greatly enhanced. Over a period of six months Louis XIII of France had forty-seven bleedings, a litre at a time. An average-sized man contains only five to six litres of blood.

Many patients were cured to death. The end of Charles II was hastened by repeated bleedings. George Washington merely caught a cold, but the extraction of three litres of blood by continual bloodletting ensured that he died from it. During the First World War some wounded soldiers were drained of even more blood.

While blood was the physician's free-flowing friend, for surgeons it was the enemy. If a surgeon was foolish enough to open a patient's chest or abdomen, every tissue his knife touched seeped or fountained opaque blood, obscuring his view and signalling the premature demise of the patient. Uncontrollable blood loss was one of the greatest impediments to the progress of surgery.

After an amputation the stump was sealed with a hot iron that fused skin, muscle and blood vessels, but by this time much of the patient's blood had already escaped. Moreover, cauterisation – or, in the case of a large bullet wound, pouring scalding oil into the hole – was often more agony than the patient could bear.

If only the blood lost could be replaced, lives could be saved. The obvious thing to do was to link up the patient to a healthy volunteer and let blood flow from one to the other. In 1650 an Englishman called Francis Potter tried direct transfusion via flexible tubes made from animal windpipes tipped with a quill to pierce the patient's vein. It was a failure. The first effective transfusion between animals was by Richard Lower in 1666. Christopher Wren, the astronomer/architect, also pioneered exchanges of blood between dogs. This led Samuel Pepys to consider: 'pretty wishes, as of blood of a Quaker to be let into an Archbishop ... but , if it takes, [it could] be of mighty use to man's health, for the amending of bad blood by borrowing from a better body'. Animal transfusions became novelty demonstrations at scientific soirées. At the Royal Society in 1667 Jean-Baptiste Denys transfused the blood of a sheep into Arthur Cogan, 'a debauched man'. Miraculously, Cogan survived, but became 'a little cracked in the head'. When one of Denys's patients died there was a scandal, until it was discovered that the man had been poisoned by his wife.

In 1829 a Doctor Blundell published a description of his 'Gravitator' apparatus for human transfusions. In his illustration the stoic donor stands to attention with his punctured arm outstretched and blood squirting into the funnel atop the Gravitator. It then flows down a vertical tube into a cannula stuck in the patient's arm. The rate of flow is regulated by a small tap and perhaps by the donor crying: 'Enough, I have but three pints left!'

Blundell claimed (erroneously) that transfusions had not 'proved fatal in any one instance' but warned that if the patient's features are 'convulsed, the flow of blood should be checked ... So long as no spasmodic twitching of the features or other alarming symptoms are observed, we may proceed without fear.' However, 'the heart and vascular system being feeble, there is reason to believe . . . that sudden death might . . . be produced'.

Wisely, he recommended that physicians should 'confine transfusions to cases in which there seems to be no hope for the patient, unless blood can be thrown into the veins'. Recently the American Society of Anaesthesiologists has issued similar advice to doctors, following studies indicating that far more infections and higher incidences of stroke, heart attack and kidney failure occur within a month of having transfusions following operations. Patients having the same operations but no transfusion fare far better.

Blundell was aware that there might be 'possible though unknown risks' in blood transfusion. How right he was. It would be over seventy years before blood groups were described by Karl Landsteiner. He wondered why so many transfusions killed the patient. When he took blood from his colleagues and mixed it with his own sometimes the red blood cells became sticky and clumped together, which would be fatal if it happened inside the body. The clumping never occurred when mixing two blood samples from the *same* person. Landsteiner hypothesised that the clumping indicated that the blood was responding to an antigen in the other blood in the mix. Antigens are substances that stimulate the body's defences to produce antibodies to neutralise them, because they are perceived as being 'foreign'.

Clearly blood would not respond to antigens that it possessed itself, so there must be different types of blood characterised by having different antigens.

Landsteiner identified two antigens, A and B. That gave four possible blood groups i.e. those with one or other of the antigens (A *or* B), or both (group AB) or neither (group O). Thus he could predict which group would react against others. Providing the donor and recipient had the same blood group all would be well, but not all other combinations were necessarily incompatible. Group O has no antigens for other groups to react against so group O blood can be given safely to anyone, whatever their blood group. It is the universal donor. Group AB already has both antigens and so doesn't perceive them as alien. AB is therefore the universal recipient. The story is a little more complicated than that as there are other factors that create incompatibility. Nonetheless, blood could now be tested in advance for compatibility before a transfusion, although this did not become routine until blood banks were established in the late 1930s. Nowadays one in three of us will have a blood transfusion at some point in our lifetime.

For transfusions it would be ideal to have large quantities of O group blood, the universal donor, but only just over forty per cent of people in western countries are type O. To solve this problem, in 1981 a biochemist called Jack Goldstein and his team deliberately transfused themselves with blood of a different type from their own.

They had found an enzyme that could snip off the antigen from group B blood cells so that they became group O. Experiments with monkeys indicated that small quantities of 'converted' B cells survived inside recipients whose immune systems would not tolerate unconverted cells. But was it safe to give to humans? There was only one way to find out. The scientists must test it on themselves.

Each participant was chosen because he had a different blood group from the others. If *all* the B group blood had been transformed to O, all would be well. If not, then some of the team, including Goldstein, would be in dire peril.

Fortunately all went well. Goldstein demonstrated that no one had developed antibodies to the converted cells. Further tests on volunteers confirmed that transformed cells could furnish a bank of O group blood for the purpose of transfusion.

Blood is a complex substance and like any other organ it is vulnerable to a variety of diseases. Many blood disorders are confined to human beings. In such cases animals do not make good guinea pigs. Self-experimentation is alive and well in many laboratories. Nowhere has the tradition been stronger than at the Washington University Medical School in St Louis or, as it is affectionately known, the Kamikaze School of Medicine.

It had a tradition of self-experimentation long before William Harrington came to the university. In 1945 he had been a student in Boston, working nights at a local hospital.

A seventeen-year-old girl came in with blood emerging from her womb. A doctor chastised her for undergoing an illegal abortion. Harrington was instructed to examine her blood and discovered that she had a dearth of platelets (blood cells essential for clotting). This was not the aftermath of a botched abortion. The girl was seriously ill, but he had no idea what disease this could be.

It was called ITP for short. We often complain that a doctor's writing is illegible but even the clearest hand wouldn't make sense out of 'idiopathic thrombocytopenic purpura'. Thrombocytopenic translates as 'deficiency of platelets', and idiopathic means 'of unknown cause'. One of its symptoms is purple bruises (purpura) arising from the slightest pressure. The prognosis was poor. Doctors had no idea what caused the condition. Frequent blood transfusions might prolong the misery, but the likely outcome was the patient would bleed to death.

The blood sample had saved the girl's honour, but not her life. She died on the operating table. Harrington decided he would search for a cure for ITP.

That was why he came to train under Carl Moore, the head of the Washington University Medical School. Harrington impressed Moore with his theory that ITP might develop when a body reacts against its own platelets. Blood cells are born in the bone marrow and end up in the spleen where they are broken down and some of their constituents are recycled. There are two obvious ways in which a catastrophic deficiency of platelets might arise: either the bone marrow ceases to produce them or the spleen goes into overdrive and destroys them faster than they can be replaced.

According to Harrington the easiest way to distinguish which of these was responsible was to inject blood from a patient with severe ITP into a healthy volunteer. If pro-duction in the marrow was to blame one would expect a gradual decline in the number of platelets in the guinea pig, but if something was destroying them his platelet count would plummet. The guinea pig would, of course, be Harrington.

Before the transfer experiment began his colleagues took marrow samples from Harrington's breastbone with a stout needle. A local anaesthetic numbed the pain, but it still felt like a stake to his heart.

At the beginning, Harrington's blood contained fifty times more platelets than that of the donor with ITP. But that was about to change. After receiving half a litre of blood from the patient, he became ill almost immediately and within hours his platelets had all but vanished. Before the day was out he was surprised that he had already developed the first signs of ITP.

The donor had received half a litre of Harrington's blood, but her platelet count didn't improve and she was bleeding profusely. It didn't augur well for her or for Harrington.

His platelets weren't increasing and his colleagues now realised what danger he was in. He bruised at the slightest touch and was terrified that he would have a stroke. Had he, a Catholic, incurred a mortal sin by inadvertently volunteering to commit suicide? Surely God would recognise his courage and selflessness.

It was several days before Harrington's platelet count began to recover. He was overjoyed, but mostly because he had shown that, although his bone marrow was normal throughout the experiment, he had developed ITP because something in the donated blood was destroying his platelets. Also, his healthy blood had not helped the sick patient because her blood had destroyed the platelets he had donated to her. Against all odds and after fifty-six blood transfusions she too fully recovered.

In the most convincing way possible William Harrington had shown that the body could turn against its own cells. It was the first clear demonstration of what we now call an autoimmune disease.

Harrington, his colleagues and technicians were the guinea pigs for many subsequent experiments to uncover the mysteries of ITP. Indeed his boss, Carl Moore, was in hospital recovering from one of these experiments when he interviewed a young chap for a fellowship. Thomas Brittingham tried to ignore the blood streaming from Moore's nose as they discussed the question of why some transfusions failed even when the blood groups of the donor and the recipient matched. Could it be that the recipient produced antibodies against alien *white* blood cells just as it did for red cells? If so, these antibodies might be a useful ally in the fight against leukaemia, a disease characterised by a huge overproduction of white cells.

Brittingham decided that the best test would be to inject himself with blood from someone with leukaemia and see what happened. What *might* happen was that he would give himself cancer of the blood. Experiments elsewhere had shown that leukaemia could be transmitted to mice and birds and even a single cancerous cell would suffice. J. B. Thiersch in Adelaide attempted to use blood and lymph from patients with chronic leukaemia, hoping that the transferred cells would establish themselves in the bodies of other patients. His hopes were dashed when they failed to contract leukaemia. He experimented on patients who were terminally ill with diabetes, syphilis, pernicious anaemia and cancers other than leukaemia. This was not, he thought, entirely satisfactory because many of them thoughtlessly died of their original illnesses without giving his experiment a chance to work. Moreover, from fighting their respective diseases they might already have had sufficient antibodies to prevent the leukaemia from taking. As they were so unsuitable it is perhaps a pity that they should have been subjected to additional stress during the final months of their lives.

Brittingham was very gung-ho about *his* self-experiment. He thought it would be great if he could prove that leukaemia was a transmissible disease. Great for science perhaps, but not for Thomas Brittingham, who was a thirty-year-old father of three.

A leukaemia patient supplied him with blood containing forty times more white cells than was normal. Brittingham strove to provide the best possible circumstances to favour infection. He repeatedly injected himself – ten times – with two large syringes of cancerous blood over a period of twenty weeks. After each injection his head pounded, he felt nauseous and was fluctuating between fever and chills for twelve hours or more. But his white-cell count did not increase appreciably as it would have had he contracted leukaemia. By his ninth injection he had developed many antibodies against the sick patient's white blood cells. He had proved his point. In the publication of his results he cautiously stated that he had no signs of leukaemia so far.

Brittingham expanded his study to include injecting himself with blood from patients with a variety of blood conditions, including cancers. One of these conditions was aplastic anaemia, in which both the red and white cells are depleted. Within moments of injecting the 'tainted' blood he felt weak and he began to eject fluids from both ends of his body. He was not only swamped externally, but also internally. His lungs were drowning and he was gasping for air. He was given oxygen to help him to breathe, but nothing seemed to alleviate the crisis.

He took a long time to recover and on the way he developed hepatitis B, got a clot in his jugular vein and, most distressing of all, he became allergic to alcohol.

It could have been worse – a nurse was spotted about to put

into his drip a type of cortisone that was for intramuscular use only. It might have killed him. It could have been worse still. At the very beginning of the experiment a friend persuaded Brittingham to inject only fifty millilitres of blood into himself – he had planned to use a dose five times greater.

Not all the self-experiments at the Kamikaze Clinic were life-threatening; some just verged on the bizarre. In 1920 Samuel Grant and Alfred Goldman studied a condition called tetany in which the body develops uncontrollable twitches that get more serious if it persists. The spasms can spread to the larynx and the spinal muscles and become far more painful.

The causes were unknown, but there was the suggestion that people having a panic attack and breathing rapidly could induce it. It called for a bout of self-experimentation. So Grant and Goldman took a deep breath – indeed, they took fourteen deep breaths a minute, in time with a metronome. Soon their fingers tingled and their facial muscles became so rigid that they couldn't speak. During one trial Goldman suddenly shrieked and his entire body went into spasm, his back arched like a bow. After twenty or so rounds of this, they concluded that hyperventilation could indeed cause tetany.

Meanwhile, 4,000 kilometres away in Cambridge University, young Jack Haldane, a super-talented Jack of all trades, was also energetically hyperventilating. He was trying to change the body's chemistry and confirm that carbon dioxide stimulates breathing. He found that over-breathing for an hour or more flushed out all the carbon dioxide from his lungs and then he felt no urgency to breathe at all. He turned blue and developed severe pins and needles in his hands. His nerve endings were still spiking two weeks later. His most striking symptom enabled him to claim the world record for one and a half hours of continual spasm of the hands and face.

Getting rid of too much carbon dioxide removes carbonic acid from the blood, making it more alkaline. Always up for a challenge, Jack decided to investigate the effects of making the blood more acid. He tried the direct method of drinking hydrochloric acid, but noted that this had a tendency to dissolve his teeth. Even a one per cent solution corroded his throat so he never cared to drink more than half a litre at a time. His calculations indicated that it would take seven litres to achieve a significant change in the acidity of his blood. He turned instead to smuggling the acid into his body 'under false pretences' by drinking ammonium chloride, which breaks down internally, liberating hydrochloric acid. The acid combines with other chemicals in the body, giving off carbon dioxide. Soon he was generating seven litres of carbon dioxide an hour. After several days of taking ammonium chloride he could hardly walk. A colleague found him 'drunk' on the stairs and rushed to his aid. 'It's nothing,' Haldane assured him. 'It's just that I'm only eighty per cent sodium haldanate at the present moment.'

To neutralise the acid and make his blood more alkaline

again he hyperventilated and swallowed eighty-five grams of bicarbonate of soda. This made his liver fizz like a 'Seidlitz' powder and led to a preoccupation with trying to breathe.

Pregnant women and their babies sometimes suffer from tetany because their blood becomes too alkaline. They are given diluted ammonium chloride and rapidly recover, thanks to Haldane's appetite for self-experimentation which, as we shall see, was far from satisfied.

Blood transfusion, 1828. Hopefully the donor's blood group was compatible with the recipient's.

A Change of Heart

'For this relief much thanks . . . I am sick at heart' — Shakespeare

The heart was many things before it was known to be the body's engine for moving blood. It was the seat of courage and love and compassion. The heart was so important to the Ancient Egyptians that it was the only organ to be left in place when they stripped out the viscera during mummification, for it was to be weighed by the gods as a measure of the man's conscience. In contrast, the Aztecs ripped out the hearts of human sacrifices to offer them to the gods, and reckless lovers offer their hearts to each other.

The heart beats out the rhythm of life, but it was not until the early nineteenth century that physicians began listening. A young doctor called René Laënnec was failing to sound the chest of an over-plump young woman and wondered if he dared to put his ear to her chest. Instead, he rolled a quire of paper into a tube and placed one end under her breast and the other to his ear. He could clearly hear the echo of her breathing and the pounding of her heart. A skilled woodturner, he made a proper listening tube that he called a stethoscope (Greek for 'spy on the chest').

It revealed that the chest was an auditorium for the internal music of the body. Laënnec was able to match the sounds he heard inside seriously ill patients with the symptoms later revealed at their autopsy. Thus he deciphered the sound of specific diseases, making the brilliantly simple stethoscope a major tool for diagnosis.

It was not universally accepted. Broussais, the lover of leech therapy, thought it was 'a useless discovery' and even into the twentieth century a few conservative doctors still preferred to use their ear pressed to the skin. For whose benefit is unclear. Were it not for the invention of the stethoscope, what on earth would medical students hang around their necks to give the impression that they are doctors?

Laënnec was a gifted pathologist and physician. He described and named peritonitis and cirrhosis of the liver. He also wrote a guidebook on using the stethoscope and, for a small extra charge, a stethoscope was included.

He contracted pulmonary tuberculosis, a disease for which a stethoscope was critical for diagnosis. He also suffered from acute angina and clinically described his own symptoms as being like 'iron nails or the claw of an animal tearing asunder the front of the chest'. Although seriously ill, he slaved over writing a major revision of his book. In doing so he displayed the same courage and dedication as the self-experimenters. 'I knew I was risking my life,' he wrote, 'but the book that I am going to publish will . . . be worth more than one man's life . . . My duty was to finish it, whatever might happen to me.' Shortly after its completion Laënnec died. He was only forty-five.

Most of the body's organs go about their business quietly with little outward sign of activity. The heart, however, is an exuberant organ. It almost shouts 'I am alive and kicking', beating 100,000 times a day for every day of your life. Its pulse accelerates by a third during pregnancy and goes far higher when you are stressed or exercising. It generates enough force to pump blood through over 199,000 kilometres (123,000 miles) of blood vessels. Perhaps it can be forgiven for failing occasionally.

No wonder the heart fascinated and frightened surgeons. It was regarded as untouchable. How could one operate on something that was never still, and with the slightest wound could explode blood outwards in an unstoppable gush? If this happened, there were only four minutes left to save the patient's life.

It was best to ignore the heart. Indeed, a highly influential surgeon of the day warned that anyone 'who tries to suture a heart wound would deserve to lose the respect of his colleagues'.

In 1896 a distinguished British surgeon suggested that

surgery had 'reached the limits set by nature ... no new method, and no new discovery, can overcome the natural difficulties that attend a wound to the heart'. Within seven years a German surgeon called Ferdinand Sauerbruch proved him wrong, albeit by accident. He had a patient with heart failure and thought that it might be because the membrane round the heart was constricting it. He managed to peel back the membrane and then, perhaps as an encore for the watching audience, he removed a cyst from the surface of the heart. It was a mistake. The cyst was in fact a protrusion of the heart's wall. He had cut directly into the heart, which responded by propelling blood everywhere. Sauerbruch calmly plugged the hole with his finger and sewed up the wound, saving the patient.

The heart is difficult to get at as it's protected within a cage of ribs. The problem lay not in cutting through the bones but in dealing with the lungs that immediately deflate when they are no longer encased in the chest. One way around this was to aerate the lungs artificially. John Hunter had used a bellows for this purpose 150 years earlier. It had enabled him to open the chest of a dog and observe the beating of its heart. Sauerbruch's solution was to enclose the patient, operating table and the entire surgical team within a large sealed chamber in which the pressure could be kept as low as that in the patient's lungs. Only the patient's head and the anaesthetist were outside the chamber. This unwieldy and expensive apparatus never caught on for chest surgery but was the prototype for the iron-lung machines that kept poliomyelitis patients alive decades later.

The problem was that surgeons were approaching the heart from the wrong direction. There *was* another route, since all the main veins of the body led to the heart. Claude Bernard, a French physiologist, had passed tubes within the blood vessels of animals, and other researchers had used his technique to measure blood pressure and the concentrations of oxygen and carbon dioxide in horses, which have conveniently wide and robust veins. No one dared to try anything so dangerous on humans.

In the late 1920s Werner Forssmann was a student studying medicine in Berlin. After the humiliation of Germany's defeat in the Great War, patriotism was on the rise. The emphasis was on 'German science'. One lecturer who dedicated himself to manufacturing German words for every medical condition called an aneurysm caused by syphilis *Hauptkörperschlagaderlustseuchenerweiterung*.

Forssmann's practical experiences were eventful. Once he assisted a doctor attending a girl who'd had a botched abortion. They operated on the rickety kitchen table. Her roommate held up the kerosene lamp but fainted, breaking the lamp and setting fire to the floor. Forssmann instantly wrenched a blanket from beneath the patient and smothered the flames. Unfortunately, the table collapsed, the patient fell off and the rest of the operation was conducted in near-darkness at floor level. But her life was saved. On another occasion, when Forssmann and his colleagues used chloroform for a delivery, the patient's husband flung himself over his prostrate wife, convinced that they had killed her. To allow the delivery to proceed, Forssmann had to lock him in the privy. He shouted 'Murder! They're killing my wife!' until she presented him with a son.

Forssmann was becoming interested in the challenge of the heart. Surgeons were developing new techniques to tackle diseases of almost every organ of the body. But if you wanted to know what was wrong with your heart, you had to wait for the post-mortem.

As a young intern he'd seen a drawing of a horse with a catheter (tube) in its neck and decided this technique would enable him to plumb the interior of the human heart. His boss would not hear of trying it out on a patient. Nor would he consider the idea of Forssmann doing it to himself.

Forssmann decided to try, and enlisted a compliant surgical nurse to secure the equipment he needed. He assured her that it was not dangerous even though he had no idea how the heart's delicate lining would take to the probing. Even slight molestation can cause the heart to exchange its regular beat for erratic tremors that in those days were invariably fatal.

During the surgeons' lunch break they sneaked into the operating theatre. Forssmann opened the vein in his arm and slowly slid sixty-five centimetres of tubing up towards his heart. They went to the X-ray room where he used a mirror to help him guide the tube into his heart. The technician took a photograph showing the tubing eerily snaking across Forssmann's body and entering the heart. It might well have been a heart-stopping moment.

The next morning he was on the carpet for disobeying orders, but his boss was impressed by the experiment and took him out to dinner. During the following weeks Forssmann repeated his self-experiment five times without mishap. He was even allowed to deliver drugs to a terminally ill patient by means of a tube along a vein.

He wrote an account of his results for publication. His boss advised him that he should say he did preliminary trials on cadavers, rather than give the impression that it was an irresponsible stunt. Around this time Forssmann moved back to Berlin to work under the great surgeon Sauerbruch.

When the article containing the authenticating X-ray photograph was published in 1929 it caused a sensation. The hospital was besieged by reporters and one of them offered a large sum for permission to publish the picture. It also caused a stir in the academic community. Some medics said that they had done exactly the same thing years before, but none of them produced a scrap of evidence to support their claims. Sauerbruch was not thrilled by the controversy. He made it clear to Forssmann that 'you might lecture in a circus about your tricks, but never in a respectable German University'. Forssmann was sacked.

He returned to his old hospital determined to take his research further. X-rays were now being used to diagnose problems in the gut by using substances that the patient swallowed to make the interior of the intestines show up on an X-ray screen. Perhaps with his catheterisation technique he could make the interior of a living heart visible.

This time Forssmann was advised to try some animal experiments first. With rabbits, the moment the tube touched the inside of the heart it stopped beating. So he switched to using dogs. The hospital had no facilities for keeping dogs so his mother housed them in her apartment. When required they were ferried to the hospital in a cab. The dogs fared better than the rabbits until he squirted a compound opaque to X-rays into their heart. Then they too died. When he switched to another contrast substance, sodium iodide, the dogs were fine and he was able to take lots of X-ray photographs and arrange them into a series to show the contractions of the heart.

But was it safe to use on humans? If rabbits and dogs reacted differently, perhaps dogs and humans might too. This was a more dangerous experiment than the original one. If that had gone wrong Forssmann could have withdrawn the tubing, but once he injected sodium iodide into his heart there was no going back. The chemical would flow throughout his body and who knew what it might do. He had dabbed some of it onto his skin and rinsed some round his mouth with no ill effect, but what did that tell him about how it might react in his heart, liver and kidneys?

Forssmann decided to push his luck by threading a tube into the vein in his neck and then down into his heart before releasing the chemical. The tube took a wrong turning, but he found the heart at the second attempt. He didn't feel well afterwards but it wasn't as bad as he feared. He made nine more insertions into the heart and not by the easiest route. The blood vessel he had used initially was now sewn up so instead he used a vein of his upper leg, pushing the tube into the main vein in the abdomen and then up into the heart. All this was done entirely by feel as he couldn't see where the probe was going.

It was more hazardous than he imagined. Years later a surgeon who avoided self-experimentation by letting his patients bear the risk instead inserted a catheter into a main artery and then into the branch leading to the kidney. Half of the patients involved suffered 'troublesome mishaps' and on one of them 'ill-advised force was used', puncturing the wall of the artery and causing severe loss of blood.

Before he was done with experimentation, Forssmann tried to inject the X-ray-opaque substance directly into his aorta (the main artery leaving the heart) using a long needle and only a local anaesthetic. The first thrust of the needle was like an electric shock. It had hit a nerve and just missed his spinal cord, sparing him from perhaps permanent paralysis. Three more failures left him exhausted and bed-bound. His wife told him that was his last self-experiment.

Sauerbruch, having fired him, had a change of heart and invited Forssmann to rejoin his team. The condescending greeting from his immediate superior set the tone: 'So you're the gentleman from the provinces who's going to teach us all about science, are you? Well, we'll see about that. First we've got to whip you into shape.' Even the word 'gentleman' sounded like an insult.

Sauerbruch was not the mentor that Forssmann hoped for. He hardly ever saw him or his patients and was given almost no operations to carry out. Nor did the great man, who had once been a pioneer himself, encourage Forssmann's research ambitions. In the rigid hierarchy everyone knew their place. Initiative was not required, obedience and hard work were mandatory. Forssmann was not getting to bed until midnight, only to begin again at six a.m. An unsympathetic surgeon thought that with six hours' sleep every night he was in danger of developing bed sores.

Sauerbruch's way of boosting morale was to regularly line up the staff and find fault with each one in turn. Some he branded 'complete idiots'. At least two of his 'idiots' went on to win the Nobel Prize. One of his extraordinary practices was to have all the patients who were to be operated on that day anaesthetised at the same time. This meant that those at the end of the list might be under for several hours before they even made it to theatre. Forssmann noted that perhaps because of this there was a high level of post-operative respiratory complications. Close attention to the patients' recovery was needed. Thus Sauerbruch pioneered intensive-care nursing, out of necessity.

With the outbreak of the Second World War Forssmann served as a surgeon in field hospitals from the opening German offensive in Poland to the disastrous invasion of Russia. Most of what we know of his wartime experiences comes from his autobiography. He tells of being offered all the facilities he could desire for his cardiological experiments in a sanatorium run by the SS. There would, of course, be an unlimited supply of patients to work on. It was one of several SS appointments he turned down. When his field hospital was about to be overrun by the Russians he simply walked back home.

After the war ended it was difficult to find a job. For a while Forssmann was banned from practising medicine until his activities during the war had been investigated. He eventually became a doctor in a quiet rural area and faded into obscurity.

In Germany his achievements were wilfully ignored. Even a history of coronary heart disease published forty years after his experiments, by which time his techniques were well established, makes no mention of Forssmann or heart catheterisation.

But surgeons abroad had not forgotten his research. One day he came across an article on the 'new cardiology' that began: 'The German Forssmann was the first...' He was delighted but felt saddened that he had not been able to exploit his pioneering technique.

In 1956 a sponsor tried to secure an honorary professorship for him at his old university, but was snubbed by a response that made it clear that an 'outsider' was unacceptable to a German faculty. Later that same year they had a change of heart and invited him to take up a chair. Perhaps because he had just been awarded the Nobel Prize.

Forssmann's experiments provided revolutionary methods for locating, diagnosing and treating heart disease. It was a quantum leap for cardiological surgeons. They had been given access to the interior of the heart without having to lift a scalpel. Conditions considered hopeless became curable. Drugs could be delivered directly into the heart instead of being injected and then diluted while travelling there in the bloodstream. The tips of the catheter were modified to scour the lining of a dangerously furred coronary artery and prevent blockage, the main cause of heart attacks. Other tubes now carry a tiny balloon that can be inflated to widen restricted vessels (angioplasty).

These techniques are now so commonplace that every patient with a heart problem is given an angiogram, an X-ray in which the blood vessels and the chambers of the heart are rendered visible, as in Forssmann's second experiment. At this moment hundreds of catheters are sliding around patients' bodies, into and through the heart to diverse destinations beyond. Every organ served by a blood vessel has been visited. Countless lives have been saved. All because of a brave man, who was fired, forgotten and finally rediscovered and rewarded when it was too late.

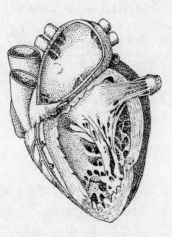

The labyrinths of the heart displayed.

Behind the Lines

'A hero may keep his head while those around him are losing theirs, but do they know something he doesn't?' – with apologies to Rudyard Kipling

It is perhaps ironic that the war finished Forssmann's research career, because wartime is a period of feverish research activity. Governments are keen to fund anything that might help the war effort or save lives.

The citation for Forssmann's Nobel Prize praised him for daring to attempt such dangerous experiments. Another example of his bravery was that he continued operating in his field hospital with Russian artillery shells exploding nearby. But courage can get you into trouble. It puts you in situations that are best avoided. There is an old Chinese proverb that says: 'Of all the thirty-six alternatives, running away is best.'

Fighting men are rightly celebrated for their bravery

under fire, but those whose courage surfaced on the Home Front are often overlooked. During the Second World War Britain was sustained by the many thousands who volunteered to become firemen, air-raid wardens, munitions workers and bomb-disposal officers.

It was not until the mass bombing of cities began that the War Ministry realised it had no idea what to do with unexploded bombs. There were no disposal squads at all. They advertised to see if anyone would like to try their hand at bomb disposal. Engineers and scientists seeking a technical challenge came forward and they were not alone. An architect volunteered, having decided that Hitler was 'becoming a bloody menace'. One of the newly trained squads comprised an earl, his secretary and his chauffeur. Their training was perfunctory. One trainee recalls that although the instructor could show them a variety of *British* bombs, he had only a single example of a German one. When asked what the markings on the casing signified, he hadn't the slightest idea. They were shown a fuse, sorted into groups and sent out to deal with live bombs.

The scale of the problem was immense. For eight terrible months London and other cities were pulverised by vast armadas of German bombers. One squadron leader said, 'There are oceans of them.' On London alone they dropped up to a thousand tonnes of bombs every night. When the air-raid siren sounded the population vanished underground or into little tin shelters half-buried in the garden. A raid might last for twelve or more hours, night after night. When a cockney woman was asked where her husband was, she replied, 'He's in the army, the bloody coward.'

One morning, when buildings had as usual been reduced to rubble, one street was found to be littered with waxen-faced bodies – Madame Tussaud's had taken a hit. Surprisingly, falling bombs were not the main cause of deaths. London's anti-aircraft guns were set to maximum elevation and banged away. The gunners admitted that they hardly ever hit a plane, but the sound of gunfire was good for morale. Unfortunately, one's morale could be dented. What goes up and explodes then falls as shrapnel. Roofs and streets clattered with a hailstorm of hot metal fragments. It is estimated that they killed far more people than the bombs.

When the siren sounded the 'All Clear' it merely indicated that bombs were no longer falling, not that they were no longer exploding. Every morning the bomb squads arrived as reliably as the milkman. Their tool kit could not be accused of over-sophistication. It contained a spanner, a brace and bit, a rope with a hook on it, a torch and mirror, a ball of string and a pickaxe and spade. Most unexploded bombs ended up deeply embedded in the ground and a large pit had to be dug to gain access to the casings and the nuts holding the bombs' fuses in place.

Unfortunately, the bomb squads had no idea what they were dealing with. One in ten bombs failed to explode, but how could the squads distinguish a genuine dud from a bomb with a delayed-action fuse? Neither alternative was a happy prospect. An indecisive 'dud' might revive at the sight of a spanner, and the bombs' time fuses were of unknown duration so the operator had no way of knowing whether they were set to go off two days hence or in two minutes' time. One officer cursorily examined a UXB (unexploded bomb) and went to collect his squad. As the men walked towards the site the bomb went off. Had they driven and got there earlier they would have all been killed.

In addition to conventional bombs the *Luftwaffe* dropped torpedo-shaped mines that were almost two and a half metres long and sixty centimetres in diameter. They descended slowly on parachutes, twirling like sycamore seeds. They were packed with TNT and the blast could fling a man a quarter of a mile. Most were on time fuses and some were magnetic. Even the proximity of a spade could set them off.

In theory a bomb could be disarmed by simply removing the detonator. In practice it was often a long and scary business. By the end of October 1940 there was a backlog of 3,000 UXBs awaiting attention.

When a bomb ended up beside the foundations of St Paul's Cathedral it was quickly dug up and loaded onto a lorry, which then raced through the East End to the Hackney Marshes. There it exploded, leaving a crater thirty metres in diameter.

Bomb disposal is so dangerous that the procedure tries to

ensure only one man at a time is at risk. The operator attempting to disarm the bomb communicates his every action to his number two who is at a safe distance. Number two writes everything down so that should there be a 'mishap' he would know what had been done right and where it went wrong. It would be his turn next and he would not care to make the same mistake.

Usually the Ammunition Examiner (as the bomb-disposal expert came to be called) worked in a deep pit kneeling in mud. Imagine the huge bomb looming over him. He strains to turn a reluctant nut. When it is free he begins to withdraw the fuse gently. He can't tell what the fuse is attached to until it is almost out. The enemy may have made the job more dangerous by adding a second fuse or linking it to a sensitive motion sensor. Rather than withdraw the fuse entirely the ammunition examiner sometimes 'trepans' the bomb, cutting a hole in the casing with a brace and bit, then using his torch and little mirror to peer into its dark interior. Like an overcautious dentist he never touches anything accidentally. He surveys the firing mechanism and forms a plan to disarm the bomb. By this time his hands are trembling and sweat is running into his eyes. They begin to burn and he can't see properly. He is alone in the loneliest place in the world.

For these men every day brought danger. It became even more perilous when the Germans began deliberately falsifying the numbers on the bomb's casing that identified what type of fuse it had. This fooled the operator into believing he was dealing with a type that he had defused several times before. In fact it was entirely different and far more dangerous. This ruse was only discovered when an examiner removed one of the new fuses and it went *phutt*. It had got wet and misfired. The lucky man had an intact fuse that could be examined in detail. That was one problem solved. Another disposal expert devised a method for using liquid nitrogen to freeze the battery in some bombs thus removing the detonator's power supply.

Someone remembered that ten years previously a German firm had tried to sell its fuse designs to the British Air Ministry. He searched the London Patent Office and found the company's patent complete with diagrams of the deadly double-capacitor circuit that was giving British bomb-disposal squads so much grief. Had anyone at the Bletchley Park code-breaking establishment riffled through the patents for encrypting devices they would have come across the details of the ENIGMA machine and could have cracked the German codes overnight.

The life expectancy of a bomb-disposal officer in the 1940s was seven to ten weeks. The volunteer earl successfully dismantled thirty-four bombs until the thirty-fifth dismantled him. The strain on the survivors began to tell. Churchill visited the squads to bolster morale. He thought they looked different from other heroes: ' . . . we are apt to overuse the word "grim". It should be reserved for the UXB Disposal Squads.'

If the examiner is close by when a bomb goes off, the blast enters every aperture of his body and he explodes outwards. At the same instant fragments of the bomb's casing become supersonic razors that shred his body. Ducking is not a viable option.

The men's greatest fear was not this instant annihilation, but the smaller accident that might leave them seriously maimed.

The George Cross was instituted as the equivalent of the Victoria Cross for those on the Home Front. The majority were awarded to bomb-disposal personnel. They were not without fear. One screamed to be hauled out of a pit because it was infested with rats.

When the war ended there were around two million tonnes of British munitions, including 130,000 tonnes of poison-gas canisters, as well as 600,000 tonnes of German ammunition, all to be disposed of. The German dumps had been partially destroyed by Allied action and the shells were scattered and damaged. This made them unpredictable and more dangerous.

High explosives were blown up. Propellants and small-arms ammunition were burnt, sixty tonnes at a time. It was like thousands of machine guns firing at once. Seventeen ships dumped the poison-gas weapons into the Atlantic Ocean.

There were numerous accidents both small and large. When thirty-seven tonnes of grenades containing nitroglycerine were detonated the explosion brought down the ceilings of houses over three kilometres away in the village of Trawsfyndd. In 1946 explosives were being loaded onto a train in Germany when suddenly there was an explosion that set off others. Two trucks and twenty-nine wagons laden with munitions simply disappeared. The disposal teams laboured through the night to douse the flames and prevent further blasts. To everyone's surprise only eight men were killed. Of all the bomb-disposal officers who died during the 1940s, a third were killed after peace was declared, while they were employed on munitions disposal.

UXBs are regularly discovered even today. They may be uncovered in a garden or on a mantelpiece. In the 1970s the television drama series *Danger UXB* alerted viewers to the fact that grandad's souvenir could be a live anti-personnel butterfly bomb filled with shrapnel. An estate agent clearing the possessions of a deceased pensioner called in the Ordnance Corps to remove his collection of grenades, shells, fuses, ammunition and mines. A long-retired squadron leader used a shell to massage his prolapsed piles (or so he said), until one day it vanished from sight. At Accident and Emergency he was attended by both nurses and a beaming bomb-disposal officer.

Anyone who tackles explosive devices knows that no matter how skilled or careful he is, sooner or later his luck may run out. To retire alive and undamaged must be an immense relief, a chance to live a life in which a sneeze is the most violent explosion you are likely to meet. Yet several of the retired officers cannot shed the need to experiment with lethal mechanisms. They spend their civilian years defusing explosive devices for the Metropolitan police.

The military bomb disarmers are now called Ammunition Technical Officers (ATOs) although the squad leader is usually referred to as 'Felix', after the cartoon cat that had nine lives. In this job nine lives might not be enough.

They are better trained and better equipped than their predecessors. They even have mobile robots to reconnoitre and sometimes remotely neutralise bombs. The less hands-on the officer is, the more likely he is to retain his hands.

Every time an ATO is killed, the device responsible is replicated to determine why it detonated. The findings are circulated so that the same mistake doesn't happen again. It is a job in which the new man truly steps in dead men's shoes.

Terrorist devices come disguised as cars, letters and shopping bags. At an ordnance depot in Northern Ireland the soldiers kept fit by playing football in the yard. The sniffer dogs joined in and worried the ball so much that it split open to reveal that it was stuffed with insulating foam and Semtex. Fortunately the foam had clogged the firing mechanism. Not every suspicious parcel lives up to expectations. One left outside Bristol Barracks was tackled with a controlled explosion that filled the air with fluttering leaflets explaining how to deal with suspicious parcels.

The task of dealing with terrorist explosive devices has become even more hazardous because they are now improvised and therefore unpredictable. They can be detonated by a time clock or a motion sensor, but also remotely by mobile phone or a car's electronic key. Sometimes the bomber watches the officer from a distance and waits until he stoops over the device. The officer may not get that far if he fails to spot the secondary device hidden in the roadside rubbish on his lonely walk towards the more obvious and prominently placed bomb.

Every ATO knows that the terrorists only have to get lucky once, whereas *he* has to be lucky every time. It takes a cool, courageous person to carry such a burden. Over twenty years of conflict in Northern Ireland ammunition technicians earned 175 awards and medals for gallantry.

Some of them do it for the thrill of working with cutting-edge technology and the intellectual satisfaction of countering the bombers' ingenuity. They know they are saving lives. They also earn the admiration of every fighting soldier, which can be summarised as: 'Bloody hell. I wouldn't do their job.'

While the disposers of bombs consider explosions as events to be avoided, one physiologist sought them out. His name was Cameron Wright.

Cam, as he was known to his friends, was one of a hidden army of back-room boffins fighting the enemy with their intellect and courage. He worked at the Royal Naval Physiological Laboratory at Alverstoke in Hampshire, an array of ramshackle wooden huts that had started life as cargo containers.

Cam carried out risky experiments on the effects of X-rays on human tissue, *his* human tissue. With the outbreak of war, the scientific adviser to the Ministry of Defence asked Cam if he would help out on 'something frightfully hush-hush'.

Barnes Wallis had designed a bomb that in theory could skip across water to explode at the base of a dam. Cam was to be the on-board observer during the trials. They didn't go smoothly. One bomb fragmented and shrapnel perforated the aircraft's elevators, another produced such a water spout when dropped that it damaged the wings. On both occasions the plane just managed to land safely.

Eventually the bomb was trained to bounce. But could it blow a hole in a dam? 'There's a small disused dam in Radnorshire,' said Wallis. 'No earthly use any more . . . and won't ever be missed. We could try and knock it down.'

So Cam found himself flying over Wales. He started the motor that set the bomb whirling on a spindle. It was the spin that enabled the round bomb to skip across the water. But the mechanism didn't release the bomb. It spun faster and faster and smoke poured from the spindle. It was primed to explode so Cam grabbed an overhead strut and suspended himself above the bomb bay, kicking the bomb. Time after time his feet shot off the spinning sphere and he lurched forward over the bomb. He later admitted to thinking, 'This is not only dangerous, but a bloody absurd way to get killed.' Suddenly the bomb fell free, leaving Cam dangling over the open bay with a dam exploding beneath him.

Explosions were Cam's speciality. One of his delights was to show newcomers around the establishment and introduce them to some of the research. He once ushered a group around a circular steel tank full of water.

'I shall now demonstrate the effects of an underwater explosion,' he announced ominously. 'Roll up the right-hand shirtsleeve and stick your arm in the water.' They did so with muted enthusiasm. 'Right. Fire!'

There was a bang and a plume of water rose from the centre of the tank. The victims assumed that their arms had been amputated at the elbow.

'That,' said Cam cheerfully, 'is what a mere 1.5 grams of explosives feels like. Now put the other arm into one of these tubes and immerse it in the tank.'

Reluctantly they slipped on metal drainpipes or foam-rubber sleeves.

'Fire!'

Those wearing drainpipes now had *two* arms that felt as if they had been blown off, whereas the rubber-clad victims suffered hardly at all. 'There you are,' said Wright. 'A pressure of 1,000 pounds per square inch and there are no harmful effects if you wear rubber. Just to prove there is no trick involved, those wearing rubber sleeves please exchange them for drainpipes and we will complete the demonstration.'

The assault divers on D-Day wore suits designed by Cam to protect them from blast. He became an authority on underwater injuries. Explosions are dangerous underwater because the incompressibility of water allows the shock wave to travel much further than in air.

A pressing task was to predict the effects of depth charges on survivors from sunken vessels. With a physicist, A. H. Bebb, Cam examined the dual deadly effects of underwater explosions: the shock wave that can shatter the diver and the 'water ram' effect that follows behind to crush him. Cam interviewed survivors of explosions at sea and repeatedly heard stories of how their companions had suddenly lost the power to move their arms and legs, given out a little gasp, then disappeared beneath the waves. He also attended post-mortems on servicemen killed in explosions. Often gross internal injuries gave no external signs.

Cam risked many an explosion. To test the possible insulating effects of an air cushion he relaxed in an inflated immersion suit whilst an enormous explosion was set off beneath him. He is said to have flown a good distance. Much of the research involved suspending volunteer navy divers, plus Cam of course, in the sea and setting off progressively larger explosive charges of TNT at ever-decreasing distances from the victim. 'It was like being hit over the head with a cricket bat,' a diver confided. After suffering about thirty blasts each, their symptoms were coolly described in a report: 'Divers showed increased pulmonary signs [lung damage] and, resulting from clinical findings in chest [broken ribs] and ears [ruptured ear drums], the decision was taken not to subject naval divers to blast at this depth.'

Cam was puzzled that large, distant explosions sometimes did more damage than ones closer to. 'Dr Cameron Wright, being somewhat disturbed on clinical grounds at the harsh effects produced – apparently by low explosion pulse parameters – dived to 50 feet at a range of 2,100 feet from a 200lb charge. After being blasted, the diver, unable to move his arms and legs, and suffering from severe pain in his back and losing consciousness, had to be pulled to the surface, on account of the shattering effect of the blast waves.' Cam was hauled out paralysed and bleeding profusely from his mouth, nose and ears.

While slowly recovering in hospital he pondered the cause of his injuries. They were far more severe than those he had previously suffered from blasts that were calculated to give a much greater shock wave. He concluded that because he was suspended in mid-water he hadn't been struck just once by the blast, but several times almost simultaneously. In relatively shallow water (thirty metres) with a rocky sea floor, the shock waves not only travelled directly through the water but were also reflected from the sea bottom to the surface and from the surface to the bottom. By ill luck Cam had been suspended at the place where all these shock waves met. As soon as he was fit enough he subjected himself again to the same experimental explosion, but at a depth at which he calculated he would miss the multiple blast. Thankfully, this time he 'did not receive any disturbing sensations'.

A few years later Cam was scheduled to lecture on the effects of underwater blast, but he was called away and a colleague stepped in. One exhibit was the chest X-ray of some poor fellow whose lungs had been shredded by an explosion. The curious colleague peeled back a label in the corner to reveal the patient's name. It was Cam Wright.

The Admiralty also wanted to know whether it was possible to escape from a disabled submarine at great depth without breathing apparatus. Attempting such a thing was a frightening prospect. Would there be sufficient oxygen in the lungs at depth to prevent the escapee from suffocating on the way up?

Wright's boss decided that the required experiments were far too dangerous, but when the man was away on holiday Cam did them anyway. He lay completely submerged in cold water in a laboratory chamber pressurised to the equivalent of ninety-one metres' depth and then exhaled while the pressure was rapidly reduced. Even to close the mouth for a moment risked having his lungs inflate and burst. The desire to breathe *in* was almost irresistible. He then repeated the experiment from 100 metres at a much slower rate of ascent (0.6 metres per second). It was described by a colleague as 'a very brave and remarkable thing to do'. As a result of his experiments the Royal Navy adopted the technique of free, buoyant ascent from depths as deep as 180 metres and Cam was awarded the Order of the British Empire for his courage while self-experimenting.

Cameron Wright was one of the forgotten heroes of the Second World War. So were many conscientious objectors. In Britain pacifists who refused to fight were interrogated at special tribunals and, if found to hold genuine humanitarian or religious convictions against killing, were allowed to choose non-combative roles as medical orderlies and ambulance drivers. These jobs were not without danger. Two of the three men to be awarded the Victoria Cross twice were not combatants but a medic and an ambulance driver,

A few 'conchies' volunteered to be human guinea pigs in medical trials. Professor Brian Maegraith tested the toxicity of anti-malarial drugs on himself, his colleagues, and volunteers from the Quaker Friends' Ambulance Unit. Kenneth Mellanby was also in the market for volunteers. He was an entomologist interested in scabies, a skin disease caused by tiny itch mites. They eat their way into the skin and mine a network of burrows, causing terrible irritation. They also deposit faeces in the burrows, which results in septic infections leading to rheumatic fever and serious problems with lymph vessels and kidneys. Some people develop a horrendous, almost incurable complication called 'crusted scabies' in which thick scabs cover the entire surface of the hands, arms and feet.

During the 1930s and 1940s scabies was a scourge in Britain. Many children suffered throughout their childhood and adolescence and were excluded from school. Towns had 'cleansing stations' for disinfecting the population. At the outbreak of war the condition reached epidemic proportions among the troops. It was estimated that the equivalent of two entire divisions of soldiers were hospitalised with scabies, not to mention the numbers of civilian workers who were also infected.

Mellanby set out to study the biology of the mites and discover how they were transmitted from person to person. He set up his research institute in a Victorian villa in Sheffield with twelve healthy conscientious objectors as his resident guinea pigs. They included a milkman, an artist, a maths teacher, a ladies' hairdresser and a winkle boiler. Every one of them was a willing participant and remained until the completion of the trails, even persuading some of their friends to participate. Although the newspapers often painted the conchies as cowardly and ignorant, all of Mellanby's group proved to be intelligent, brave and blessed with a sense of humour. They decided the team needed a coat of arms. Perhaps with a yellow streak, one of them suggested. They chose as their motto 'Itch dien'.

To see how easy it was to acquire the mites they slept in bedding from infested patients and even wore their underclothes. Not one of them caught the infection. They felt cheated.

At a lecture to military officers Mellanby stated that scabies was contracted by picking up a young adult female. The audience collapsed with laughter. He meant a female *mite* — although, when he thought about it, he reckoned that either type of female might do the trick. Perhaps it *was* a sexually transmitted disease. The mites certainly burrowed happily into the skin of the penis. Mellanby wondered if infected women could be hired to sleep with the volunteers. Would the volunteers be drawn to a scabious corpus? Would experimental adultery look good in the scientific report?

Fortunately, before any women were enlisted two volunteers became infected. The symptoms of scabies took longer than a few days to appear after infection by the mites; the incubation period was up to two months.

The infected volunteers subsequently shared platonic beds with the uninfected ones. Even without intimate contact, a warm bed resulted in successful infection. To follow the progress of the disease they endured the misery it inflicted for nine to eighteen months. They scratched their pyjamas to shreds. Despite suffering from unpleasant infections no one asked for the trial to be curtailed. To test the influence of cleanliness half of them bathed regularly and the others not at all. Shunning washing did not make the symptoms worse although it did reduce the doctors' enthusiasm for examining the volunteers.

They disproved several myths about the spread of scabies. Received opinion was that verminous soldiers carried the infection into the family home while on leave. In fact it was the other way round. Soldiers were regularly examined and treated, but were reinfected on visits home. The volunteers wrote a poem to clarify this:

Recondite research on a mite
Has revealed that infections begin
On leave with your wife and your children
Or when you are living in sin.
Except in the case of the clergy,
Who accomplish remarkable feats,
And catch scabies and crabs
From door handles and cabs,
And from blankets and lavatory seats.

A hospital unit was set up to test various 'cures' for scabies. It treated as many as 150 patients per week.

Scabies was only one of the volunteers' ordeals. With such willing guinea pigs on hand, Mellanby asked if they would join in other trials. They all volunteered for dietary experiments. As men of principle, they could be trusted absolutely to stick to a diet no matter how unpalatable. They tested the digestibility of the new National Wheatmeal Loaf, whose constituents, it was feared, might impair the body's ability to absorb calcium. The fears proved to be unfounded.

They also studied vitamin deficiencies. The volunteers survived for almost two years on dull and deficient diets. Those deprived of vitamin C developed scurvy and were treated to wound-healing experiments in which cuts made into the thigh were allowed to heal and then weights were used to measure the force needed to reopen the wound.

Other experiments involved 'surgical shock'. This is not the immediate shock of being injured but the trauma that develops later and manifests itself as sweating, plummeting blood pressure, rapid pulse and collapse. The volunteers were injected with a chemical to induce all these symptoms and they were then subjected to many 'somewhat alarming experiences'. A thoroughly bad time was had by all.

To evaluate new anti-malarial drugs they were given a virulent, potentially fatal strain of the disease. All the volunteers became 'unpleasantly ill'. Because of the potential danger the researchers decided to curtail the experiment, but the volunteers insisted that, because it was designed to relieve human suffering, they too should suffer for as long as the experiment required.

You didn't have to be a conscientious objector to avoid military service. Those employed on essential war work were also excused. They included scientists working for commercial companies creating new pharmaceutical drugs. During wartime there was an urgent demand for the newly available antibiotics and other drugs. The staff at a Danish pharmaceutical company called Medicinalco, encouraged by the head of research, became enthusiastic self-experimenters. Like a band of brothers they risked their health to hasten drugs to market. They called themselves 'The Death Battalion'.

There was no compulsion to take part, nor much reward for doing so. They might get a glass of port for giving blood samples. At their annual dinner those who had taken the most risks or suffered the worst side effects were awarded a small plastic skeleton as a *memento mori*. They were worn with as much pride as any military medal.

Years before, von Pettenkofer had likened his quaffing a cordial of cholera to the act of a soldier: 'I would have looked death quietly in the eye . . . I would have died in the service of science like a soldier on the field of honour.'

Suffer

'Certain gassing now, but maybe live to tell the tale' – John Scott Haldane, enticing volunteers for experiments on poison gases

On the battlefield, when the blood is hot and charged with testosterone, many men may attempt something impulsively brave. In contrast, the calm, calculated courage of the self-experimenter is a more considered form of daring.

Jack Haldane had a surfeit of both hot and cool courage. He loved the war. April 1915, one of the happiest months of his life, was spent under constant bombardment. He relished night-time forays into no man's land to eavesdrop on enemy troops or lob a bomb into their trench. Field Marshal Haig called him 'the bravest and dirtiest officer in the army'.

He often spiced bravery with bravado. In daylight, he cycled across open ground in full view of the Germans in the belief, fortunately correct, that they would be too surprised to open fire before he made cover. He called it 'taking a novel risk, which you are not ordered to take . . . and enjoying it'.

Jack wrote home that he'd got a 'ripping job' as a bomb officer. He made trainees attach a detonator to a fuse with their teeth. He forewarned them that if the detonator exploded their mouth would be considerably enlarged. Another drill was to play catch with lighted bombs before throwing them. He struck fear into his fellow officers when he tamped down his glowing pipe with a detonator while lecturing them on how easily accidents can happen.

Jack was trained in 'the practice of courage' by his father John Scott Haldane, a medical man who became a famous physiologist. Haldane took Jack down a mine when he was a small child and not surprisingly the boy was frightened. On a later trip young Jack had to descend the main shaft in darkness, leaping from one descending ladder to another. They got lost in a maze of galleries and crawled into a shaft contaminated with methane. Jack was told to stand up and recite the 'Friends, Romans and countrymen' speech. Within a few moments he collapsed into the breathable air below. Thus Jack learned that methane was lighter than air and not lethal, at least not in the short term.

The influence of air quality on human health fascinated John Scott Haldane. He analysed the air in slum housing, factories and sewers – a comparison in which sewers came out better than schools. It was not without risk. To collect air samples he climbed down a shaft at a sewage works in which, only hours before, five workers had been killed by hydrogen sulphide gas.

Later he stalked the London Underground, sucking up air samples with a tube. The levels of deadly carbon monoxide were so high that his findings led to the electrification of the lines.

The best place to find deadly gases was in mines. Underground explosions were commonplace and after every disaster Haldane snatched up his mining helmet and rushed off to investigate. To reassure his wife, he sent telegrams that were so incoherent they merely confirmed that he was suffering from exposure to one poisonous gas or another.

He could recognise most gases by their taste. Following an explosion at one mine he identified the contamination by sucking undiluted gas from the pipe that vented noxious gases from the mine. He immediately began to pant and his face turned blue. 'Carbonic acid,' he declared. Then he sucked the pipe twice more to confirm his diagnosis.

Haldane found that most of the fatalities in mine explosions resulted not from the blast, as everyone assumed, but from suffocation because of lack of oxygen or exposure to carbon monoxide. To test the effects of carbon monoxide poisoning he inhaled it whilst recording his symptoms and taking samples of his own blood for analysis. When he could no longer stand the experiment terminated and so almost did he. The carbon monoxide saturation in his blood was only four per cent lower than that found in asphyxiated miners.

He was able to take such a risk because of his fellow experimenter, a mouse. Although Haldane advised others never to use an animal if a man would do, in this case it was essential. The rapid breathing rate of the tiny creature meant that the exchange between air and blood was around twenty times faster in the mouse's lungs compared to a man's. The rodent should therefore be twenty times more susceptible to poisonous gases.

In their experiment they shared the same toxic mix of air and carbon monoxide. Within a minute and a half the mouse was in difficulties and was therefore rescued to recover. Haldane persisted and after half an hour he developed the same symptoms as the mouse. They had taken twenty times longer to appear. A small bird with its higher metabolic rate would be even more sensitive than a mouse.

This research led to the introduction of canaries into mines as an early-warning system for bad air. Their claws were trimmed to ensure that when they passed out they also toppled from their perch, thus making their plight obvious. Haldane also designed the canary's cage. As soon as the bird swooned, the sides of the cage could be sealed to make a box and the carrying handle contained a tiny cylinder of oxygen to revive the canary. It also recovered twenty times faster than a man.

To demonstrate that coal dust was responsible for most

underground explosions, an experimental 'gallery' thirty metres long was fabricated on the surface from large boilers welded end to end. The ledges inside were powdered with coal dust and a small charge was detonated at one end. The explosion shot along the pipe and tore the last two boilers to bits. Haldane and his son stood over 300 metres away, but a huge sheet of metal flew over their heads. The explosion was heard ten kilometres away. Later experiments showed that limestone dust could inhibit explosions.

Haldane also showed conclusively that pneumoconiosis resulted from the inhalation of dust. Practical problems such as these were the stimulus for almost all of his physiological work and the results greatly reduced the risks in several hazardous professions.

Haldane's brother Richard served on the Explosives Committee at the War Office and gave a talk advertised as 'A public lecture on explosives by Mr R. B. Haldane MP, illustrated by experiments conducted by Professor J. S. Haldane'. The police saw the poster and arrived early to clear the front three rows before the professor's explosions cleared them more dramatically.

In the attic of his home Haldane had kitted out a laboratory with an airtight chamber so that he could investigate the effects of various gases. He sometimes enlisted his daughter Naomi to keep an eye on him and, if he collapsed, to flush out the poison gas, drag him out of the chamber and perform artificial resuscitation. She was twelve at the time.

Haldane was an eccentric with an unpruned moustache exploding from his upper lip. He became the epitome of the absent-minded professor. After working throughout the night, he rose in time to have lunch for breakfast. On one occasion he arrived late for dinner, having forgotten they were expecting guests. He shot upstairs to change, but didn't return. His wife went to see where he had got to and found him asleep. 'I suddenly found myself taking my clothes off,' he explained, 'so I thought it must be time for bed.'

In 1906 he turned his attention to the effects of high pressure when asked to investigate the physiology of deep diving. The Admiralty was concerned that many divers were coming up unconscious or paralysed with the bends. After some calculations and a few tests on goats, Haldane sent volunteer navy divers down to almost double the maximum permitted depth of thirty metres. Thereafter the world depth record was broken time after time. He produced the first set of decompression tables to get divers safely back to the surface and established the 'Haldane principles' on which all subsequent tables were based. Haldane had determined that a diver could rise to half the maximum depth of the dive without ill effects. Thereafter he must ascend towards the surface in stages, stopping at prescribed depths to breathe away the dangerous nitrogen he had accumulated while below under higher pressure.

In April 1915 the Germans launched their first attacks using poison gas. They released a green cloud of 168 tonnes of deadly chlorine that drifted towards the Allied positions as fast as a man could run. Many were blinded, turned blue and writhed in agony. 'Glue' poured from their mouths. Chlorine strips the lining from the lungs and turns it into mucus that blocks the windpipe and fills the lungs. Some soldiers died in their trenches. Fields of browned grass were littered with dead cows.

Richard Haldane, now the Lord Chancellor, asked his brother to identify the poison gases being used by the Germans and devise some protection for the troops. Haldane rushed to France to attend the post-mortem of a victim of the first gas attack. He immediately diagnosed chlorine inhalation, because he'd had 'an alarming personal experience in a sewer connected with bleaching works'.

Kitchener entreated British mothers to make gas masks for the troops out of stockinet and cotton wool. Was it just a ploy to distract an anxious populace and make them feel that they were doing something to help? Ninety thousand of these home-made 'gas masks' were dispatched to the front. They were worn by the 2nd Lancashire Fusiliers when they were subjected to a gas attack. Afterwards it was reported that the Lancs fusiliers had 'ceased to exist for military purposes'.

Meanwhile, the Haldane house echoed with the sound of coughing and retching from the attic, signifying that the experiments were progressing well. Haldane and two collaborating chemists were testing prototype respirators in the sealed chamber filled with deadly chlorine gas. Even chlorine at a concentration of only 0.1per cent made it practically impossible to take a breath and irritation of the eyes was excessive. Haldane's daughter Naomi and their lodger Aldous Huxley shredded woollens to provide the absorbent filling for experimental respirators. They tried stockings, vests, Naomi's knitted cap and Aldous's scarf. They also raided the kitchen for chemical absorbents.

In the meantime soldiers were dying from gas attacks so Haldane came up with makeshift devices that could be improvised by soldiers in the field. If they breathed through a handkerchief filled with soil or a bottle with the bottom knocked out and filled with damp rags, these primitive filters gave some protection to the lungs. The eyes could be shielded with gauze dipped in linseed oil without completely obscuring the soldier's vision.

Haldane also set up a laboratory in France with an experimental chamber. The guinea pigs were Haldane, an associate from his mine-research days who was a conscientious objector, and Jack, temporarily seconded from his unit. They tested the effects of various concentrations of chlorine gas with and without a respirator. The chamber contained an exercise wheel to ensure that the respirator worked while the wearer was running away. Jack later explained the drill: 'As each of us got sufficiently affected by gas to render his lungs duly irritable, another took his place . . . Some had to go to bed for a few days, and I was very short of breath and incapable of running for a month or so.' Despite his weakness, Jack was called back to active duty. En route to the front he was badly wounded by an explosion. It saved his life, for in the following few days almost every officer in his battalion was killed.

They eventually came up with an effective respirator. The only recognition awarded to anyone involved was a Military Cross for the brave young adjutant who opened the car door for the general when he visited the laboratory.

Soon the gas masks were being manufactured at the rate of 70,000 per day, although at first the factory accidentally used caustic soda instead of washing soda as the absorbent.

Naomi believed that her father's lungs never fully recovered from his poison-gas tasting sessions. At the age of seventy-five he collapsed and developed pneumonia. Haldane was confined to an oxygen tent, a device he had invented. He pioneered the use of oxygen to relieve damaged lungs. John Scott Haldane passed away with a look of intense interest, as though monitoring a critical experiment in physiology.

Three months after his death *The Times* announced that he was to give a public lecture. It was entitled 'Miracles'.

Haldane preferred to experiment on himself or others who were sufficiently interested in the work to ignore pain and fear, much as a soldier would 'risk his life or endure wounds in order to gain victory'. The Haldane family motto was 'Suffer'. When young Naomi fell heavily and cried, her father made it clear that this was 'strictly forbidden in the code of courage'. He was not inordinately brave. He had no head for heights and was too nervous to learn to swim, but the combination of scientific curiosity and the desire to help his fellow men generated courage. And it was contagious.

Even as a child Jack donated blood for his father's experiments and cajoled his school chums to do the same. Young Jack was intellectually precocious. On an expedition his father realised that they had forgotten to bring log tables. 'Never mind,' he said, 'Jack will calculate a set for us.'

He became one of the most influential biologists of his day. He laid the foundations of both human genetics and population genetics. By marrying genetics with Darwinian natural selection he also created modern evolutionary biology. In 1938 Jack was elected a Fellow of the Royal Society, as his father had been. He was forty-six years old, bald, with twinkling eyes and the air of a mischievous walrus.

He turned his lively mind to the problems of the looming conflict with Nazi Germany. In Madrid during the Spanish Civil War he had spent his time minutely recording the results of air raids, noting what gave protection and what did not. He published an article on the mathematics of air-raid protection, as well as a practical guide to air-raid precautions that sold well. He tried in vain to persuade the British government of the value of deep shelters. To prove the inadequacy of the flimsy above-ground Anderson shelters, he offered to sit in one while explosives were detonated close by. Jack condemned the government's policy as leaving London unprotected. Fortunately, when air raids began

Londoners used their initiative by breaking into the Underground stations to use them as deep shelters.

Even before war had broken out, Jack had alerted the British government to the potentially destructive power of atomic bombs and later he played a major role in evaluating the genetic damage caused by radiation. He also put forward a fanciful scheme for the release of thousands of fish tagged with tiny magnets to trigger magnetic mines.

His moment came three months before war erupted, when the submarine HMS *Thetis* sank during trials in Liverpool Bay, having dived with both ends of its torpedo tubes open. Though the vessel was intact with its stern protruding from the water, only four of the hundred and eight men on board survived. Rightly fearing that there was no help available at the surface, the crew left it too late to escape. Almost half the victims were civilian mechanics and the trade unions asked Jack to represent their interests at the public inquiry.

With a couple of colleagues he locked himself in a pressurised steel chamber to simulate the effects of incarceration in a disabled submarine. They remained there for fourteen and a half hours and, as the carbon dioxide concentration rose, became too sick and incapacitated to put on the Davis submarine-escape gear.

It brought home to Jack the terror of being trapped underwater. He confided to his sister how terrible it must have been for men trapped in the *Thetis*'s escape chamber with the water rising, unable to get the hatch open. He thought it would be advisable for physiologists to research the effect of abnormal conditions on people *before* they were killed by them.

He convinced the Admiralty that to increase the chances of submariners surviving in a disabled submarine he needed to investigate how people reacted when breathing a deteriorating atmosphere under pressure. Thus he became one of the few card-carrying Communists to carry out secret research for the War Office.

He enlisted four members of the International Brigades (because they would be cool under pressure), his secretary, and a research student who later became his wife. All of them were tested, if not to destruction, then at least to unconsciousness. Almost every experiment ended with someone having a seizure, bleeding or vomiting. 'Good,' Jack would say, 'that's another point on the graph.' Nosebleeds were so common that you could usually track him down by following the trail of bloody cotton wool.

The experiments were conducted in a 'pressure pot', a steel chamber like a boiler on its side, measuring 2.4 metres long by 1.2 metres in diameter. Two or three people could squeeze in, but they couldn't begin to stand up. The guinea pigs had no lamp or telephone. They communicated with the exterior by tapping in code or holding up messages to the tiny porthole.

Jack described what it was like to be in the chamber:

'I am breathing rapidly and deeply and my pulse is 110 . . . my writing is a little wobbly. But why cannot my companion behave himself? He is making silly jokes and trying to sing. His lips are rather purple . . . I feel quite unaffected; in fact I have just thought of a very funny story. It is true I can't stand without support. My companion suggests some oxygen from the cylinder . . . To humour him I take a few breaths. The result is startling. The electric light becomes so much brighter that I fear the fuse will melt. The noise of the pump increases fourfold. My note-book, which should have contained records of my pulse rate, turns out to be filled with the often repeated but seldom legible statement that I am feeling *much* better, and remarks about my colleague, of which the least libellous is that he is drunk. I put down the oxygen tube and relapse into a not unpleasant state of mental confusion.'

A much larger chamber could be flooded to a depth of over two metres so that strenuous tests could be carried out underwater. High pressure and cold water were a bad combination. When clad only in shirt and slacks, immersed in a bath of melting ice and breathing an aberrant atmosphere at a pressure ten times greater than normal, unconsciousness came mercifully soon. Even hardened divers suffered severe claustrophobia in this tank. Jack admitted: 'It was a queer experience to wait underwater in the dark tank, knowing that one might lose consciousness any moment, and perhaps wake up with a broken back, conceivably not wake up at all.'

His father called the pressure pot 'the chamber of horrors'. It is easy to see why. During compression it became hot in the chamber, in the same way as a bicycle pump warms when you inflate the tyres. Jack fanned himself with a folded newspaper, but it tore to shreds in air so dense that a bluebottle was unable to fly. When decompressing, the cooling damp air filled the chamber with fog. After a few weeks Jack's wristwatch ground to a halt under the weight of rust on the mainspring. His wife bought him an airtight watch. It crumpled the first time he compressed.

The speed at which they changed pressure sometimes caused problems. Jack's fastest 'dive' was from one to seven atmospheres in ninety seconds, during which he experienced the pressure changes of a pilot diving vertically at twice the speed of sound. Rapid 'ascents' were more dangerous, causing one of his filled teeth to emit a high-pitched scream and explode because of air pockets that couldn't vent fast enough. A colleague had his right lung collapse on several occasions. If both lungs had gone, he would have died. Thanks to his lung problem he failed the army medical and, ribbed by his colleagues, he was able to sit out the war in the 'comfort' of the chamber of horrors.

Minor bends were commonplace. Haldane was partially paralysed in his left buttock, but felt fortunate that 'it wasn't in a more important sensory region'. He also burst both his eardrums. They healed, leaving small holes in the membranes. It made him slightly deaf, a small price to pay for the ability to blow smoke rings through his ears.

At that time working at pressures equivalent to depths of sixty metres called for exceptional skill and technical capability. The simulated dives to 120 metres achieved by Jack's team approached the lowest depths then conceivable. The research changed the procedures for escaping from submarines and established sixty-six metres as the lower limit for the safe use of compressed air whether escaping or diving.

Wartime continually threw up new problems and created an urgency to solve them. Frogmen and underwater Charioteers (manned-torpedo operators) used oxygen re-breathing apparatus because it didn't give off tell-tale bubbles, so the team investigated the effects of oxygen under pressure. This required one of the most exhaustive programmes of diving experiments ever attempted – more than a thousand 'dives' in toxic conditions. Oxygen, the gas that supports all life on the planet, is poisonous under pressure, causing nausea, paralysis and convulsions violent enough to break a bone. Jack suffered crushed vertebrae and a dislocated hip from a sudden muscular contraction. Of particular danger was the unpredictability of the seizure. On one occasion the same volunteer lasted eighty-five minutes at a pressure of three atmospheres, another time only thirteen minutes. They established that breathing pure oxygen was dangerous below eighteen metres.

After about a hundred experiments, Jack became so sensitive that he began to twitch violently after only five minutes' exposure to oxygen. These tests sealed in a pressurised 'coffin' even got to fearless Jack. He had nightmares in which he couldn't escape from the chamber. But it didn't stop him from doing it again.

When the Admiralty introduced miniature submarines, they asked Jack to see if it was possible to survive a long journey submerged in such a tiny vessel and for a diver to leave it underwater to plant a mine on a ship's hull and then re-enter the sub. Despite his distaste for submarines, Jack and his stalwart collaborator Martin Case incarcerated themselves in a mock mini-sub made of steel and placed on the bottom of Portsmouth harbour, sealed in like 'shrimp paste in a jar'. As the tank was being lowered towards the water the air-raid siren went, and so did the crane driver, leaving them dangling in mid-air with bombs falling all around. They stayed locked in the submerged 'sub' for two days with the light and telephone links working only intermittently. A ship passed too close overhead and the tank was torn from its moorings, shaking up the prisoners inside. They concluded that with one cylinder of oxygen the crew could live comfortably for three days and stay submerged for twelve hours without any refreshing of the air. Jack's idea of comfort was, of course, not quite the same as yours or mine. Later he exposed himself to a pressure of ten atmospheres and then plunged into water at zero degrees Celsius as a rehearsal for going for a sub-sea stroll in Arctic waters. These experiments paved the way for the successful raid on the battleship *Tirpitz*.

In 1943 the War Office anticipated that, come D-Day, frogmen would be expected to clear mines and shore defences. The divers might have to surface quickly to avoid underwater explosions. If they breathed a mixture of air and oxygen they could come up faster in safety. The trick was to get the proportions right: too much air and they would get the bends, too much oxygen and they would suffer fits. So Jack and his wife were compressed to the equivalent of twenty-one metres down to try out various gas mixtures. If either of them got the bends, they gave themselves a day off. They showed that a surfacing procedure, which according to the tables should take forty-seven minutes, could be done safely in two. This gas mixture was used in 1944 by a hundred and twenty frogmen who cleared the sea of obstacles the day before the D-Day landings.

Other contributions to the war effort included reviewing all the recently published scientific work from Germany and advising the Royal Air Force on bomb-aiming devices. Jack also carried out many statistical studies on topics such as the interpretation of casualty figures and the best tactics for shooting down V-1 rockets. His only honour for all his war work was to be included in the Nazis' list of those to be arrested when England was invaded.

During the war he strolled into a pub in Gosport dressed as a Nazi officer accompanied by Cam Wright, the expert on underwater explosions, in full diving dress including leaden boots and copper helmet. No one turned a hair.

In 1964 he was diagnosed as having a malignant tumour. He wrote a poem called 'Cancer's a funny thing'. It begins:

> I wish I had the voice of Homer
> To sing of rectal carcinoma . . .

I read his poem for the first time while in hospital waiting for an operation on *my* rectal carcinoma. It was what Jack would have called 'a singular and interesting experience'. I was luckier than he was. Within a few months of his operation science had lost one of its most gifted and courageous spirits. Both he and his father left their bodies for medical research and teaching. After all, they had been used for those purposes during their lifetime.

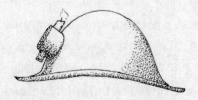

John Scott Haldane's candle hat — for detecting explosive gases?

Adrift and Alone

'Water, water everywhere, nor any drop to drink' – Samuel Taylor
Coleridge

Jack Haldane was by no means the only biologist drafted
in to do special war work. He was not even the only Jack.
In the early 1930s young Jack Kitching, newly graduated
from Cambridge, toughened himself by taking up underwater ecology off the coast of Argyll. His diving equipment was a milk churn with a glass window, a garden hose
and two car-tyre pumps. He could easily have been
mistaken for an itinerant ironmonger. He carried less fat
than a lollipop stick and his 'diving suit' consisted of a
rugby shirt, long shorts and plimsolls. With such meagre
insulation it was difficult to withstand the cold Atlantic
for more than twenty minutes at a time. On shore Kitching
was seized by uncontrollable shivering. After half an hour
in the warm Scottish rain he went underwater again. These

forays would prepare him for taking on the entire Atlantic Ocean.

In 1942 a team of scientists and engineers was being put together in Canada to tackle the physiological problems faced by airmen. Kitching was sent to the Department of Medical Research at the University of Toronto to work under the auspices of the Committee on Aviation Medical Research. Over the next three years he would write sixty-five reports on fifteen different topics. The projects were stimulated by problems encountered by the Royal Canadian Air Force. Some investigations were tackled in collaboration with scientists affiliated to the Royal Air Force or the US Air Force.

The researchers were asked to prevent the pilot blacking out when pulling out of a dive. During such a manoeuvre his blood rushed downwards away from his brain. The team developed water-filled flying suits in which the water rushed down to the pilot's legs, compressing them and keeping the blood in his head.

On reconnaissance missions the planes either flew very low, or high leaving tell-tale contrails behind them. Either way they were easy to spot and shoot down. If they could fly even higher they would be out of the range of anti-aircraft batteries. Unfortunately, at 12,000 metres the air is so thin there is insufficient oxygen to sustain the crew. Mount Everest is only 8,845 metres high but, if instantaneously whisked to its summit, I would rapidly collapse from lack of oxygen and would die unless promptly carried down the mountain. Mountaineers attempting major peaks take care to acclimatise at altitude to enrich the blood's oxygen-carrying capacity, and use oxygen respirators as they approach the summit. Above 8,000 metres is known to climbers as 'the death zone'.

The fuselages of military aircraft were not airtight or pressurised. If they were, the smallest piece of shrapnel that punctured a fuselage would result in instant decompression. As paying passengers we pay little attention to the stewardess's announcement that: 'In the event of a loss of cabin pressure oxygen masks will drop from the ceiling.' She never elaborates on 'the event'. There is no mention of a refund for passengers sucked out of a broken window. She also fails to mention that if we don't breathe oxygen within thirty seconds of decompression, we become unconscious and comatose. The pilot has only fifteen seconds of 'useful time' in which to take corrective action – i.e. put the plane into a steep dive to lose altitude.

The aircrew of fighters and bombers had to be given a personal supply of oxygen so the Toronto team investigated the problems with oxygen respirators. In normal breathing we have to suck air in by expanding the chest. Breathing out is passive. In contrast, pressurised oxygen from a respirator fills the lungs and has to be forcibly expelled by the aviator against the pressure. The trick is to have the oxygen pressure sufficient to sustain the man but not so high as to exhaust him when he tries to breathe out.

Jack Kitching, in collaboration with RAF researchers, also studied brain function and the physiological effects of breathing pure oxygen, such as nausea and impaired vision, and trialled a drug as a temporary antidote for some of the problems. He did this by self-experimentation in a decompression chamber not unlike that used by Jack Haldane. Within months a greatly improved respirator was in use with crews of high-flying reconnaissance planes.

Respirators were fine while the crew were inside the aircraft – but what would happen when they had to disconnect to bail out? The Physiological Research Centre in Farnborough was the RAF's equivalent of the Canadian group. It was tasked with determining the maximum altitude from which someone could parachute without breathing apparatus and survive. Edgar Pask, who had been an anaesthetist in Oxford, volunteered for the job. It involved dangling from a scaffold in a parachute harness while breathing air mixtures that contained less and less oxygen. At the lowest concentrations tested (equivalent to those found at an altitude of 12,000 metres) Pask's muscles twitched violently. He also had severe difficulty in breathing and passed out. Had the experiment not been curtailed he would have suffocated. This dangerous experiment established that a flyer parachuting from no higher than 10,500 metres would have a reasonable chance of surviving. Above that an oxygen supply was essential.

Bombers flew at altitudes of between 7,600 and 10,700

metres. The temperature outside was -30°C to -40°C. The 'waist gunners' in B-17 and B-24 bombers sat in plastic blisters on the side of the fuselage. The machine gun protruded through an open port allowing freezing air to swirl into the plane. The gunner's thick gloves made it almost impossible to manipulate the gun. If he took them off, his hand froze to the cold metal. Pulling the hand free left the skin behind.

Self-experiments in a cold room 'under conditions of extreme discomfort' revealed reduced reaction times that might have serious consequences if the pilot were controlling an aircraft. The critical temperature to maintain manual dexterity was 12°C.

Kitching also tackled the problem of keeping aviators warm. The thick insulation needed for clothing to combat such low temperatures was incompatible with manual dexterity. The pilot was either too cold or too encumbered to do his job. Simple innovations were curved gloves and flying suits with bent knees. It was far more comfortable to have the clothing modelled to the posture that the man adopted in the plane. Kitching was the prime mover in the development of electrically heated gloves, flying suits and inner boots that were far warmer and less bulky.

If cold was a problem in the cockpit, it was even more serious for the aviator who had ditched in the sea – providing he didn't drown first. When Japan entered the war, the supply of kapok, the buoyant material in life jackets, was cut off. To find a replacement the Toronto team tested the buoyancy of numerous materials before settling on 'milk-weed floss'.

Life jackets were supposed to prevent the wearer from drowning. At least, that was the theory. While serving as an observer on a rescue launch Edgar Pask was appalled at the number of corpses that were wearing life jackets and floating face down in the water. Designers came up with several new types of 'Mae West', so-called after the voluptuous film star who looked as if she might have an inflatable secreted in her corsetry. All the new jackets were supposed to force the wearer onto his back.

Testing their self-righting capability should have been easy. A volunteer would lie face down in a pool and leave the jacket to do its stuff. It was, however, psychologically impossible to remain inert in such circumstances. The guinea pig would *have* to be unconscious. So Pask donned a flying suit and prostrated himself in the water. He breathed through a tube in his windpipe attached to a poolside anaesthetising machine. An inflatable cuff around the pipe prevented him from swallowing water. A preliminary trial showed that without a Mae West he sank to the bottom. Dozens of tests were needed to determine the design of jacket that ensured the 'survivor' would float face up. To check that the the jackets would self-right in a choppy sea the team moved to the Elstree film studios, which had a large tank with a wave-making machine for filming model boats on stormy 'seas'. The life jacket performed well in metre-high waves. The unconscious Pask could easily have drowned in these tests had water seeped past the inflatable cuff and into his lungs.

In 1943 there wasn't even an agreed technique for resuscitating a drowning man who had stopped breathing. Different methods were promoted by four agencies. The Farnborough Centre put them all to the test.

In the absence of recently drowned volunteers, Pask accepted the challenge. It involved a hazardous experiment. A tube in his windpipe would measure how much air was forced into his lungs by the artificial resuscitation, but Pask had to stop breathing before he could (hopefully) be revived. He was repeatedly anaesthetised to the point of respiratory arrest. Unless his breathing could be restarted, and fast, he was a dead man. The experiments took two sessions of over four hours each before they decided which was the best method. Pask's heart stopped sixteen times. I know of no one who has deliberately induced so many near-death episodes on behalf of his fellow men. Pask merely claimed that he had done all his research while asleep.

Though drowning may have been averted for the ditched aviator, hypothermia would soon take hold. We are very susceptible to cold. I recently read of a woman who collapsed from hypothermia in a supermarket. She had stolen a frozen chicken and hidden it under her fur hat.

Seawater conducts heat away from the body twenty-five times faster than air does. Immersion in water below 20°C leads inexorably to hypothermia and death. In the North

Atlantic in winter a 'survivor' lasts only thirty minutes. Even when sitting on a life raft off Florida on a warm summer night, men exposed to sea spray became severely chilled. The insulation in a flying suit is useless when wet.

In October 1943 there was an urgent request for a water-proof 'ditching suit' to protect against immersion. Within a month, a butyl survival suit resembling a limp blimp was being tested in an ice-cold pool with an air temperature of 0°C. Kitching lasted four hours in 'relative comfort'.

He was also one of the volunteers for the sea trials. These involved being abandoned off Nova Scotia in November in a tiny rubber dinghy. It was supplied with a 'help' kit containing hooks, line and sinkers as well as darning needles and thread. This ensured there were plenty of sharp points around to puncture the inflatable boat. There was also a fetching 'mosquito head-net' and matches should Kitching need to light a fire. The survival rations included pea-soup powder which was 'unappetising' and chocolate that was 'extremely nauseating'. This was serious criticism from a man who was fearless in the face of food. I once saw Kitching consume a large chocolate bar writhing with maggots and his only comment was 'full of protein'.

Later sea trials of an improved survival suit demonstrated that it kept the water out during fifteen minutes' immersion followed by seven hours' exposure on a raft. By the end of the trial those *not* wearing the suit were 'in a very poor condition'. The survival suit went into mass production and was soon being issued to bomber crews. In 1947 Jack was awarded the Order of the British Empire – 'For **O**rdinary **B**loody **E**ffort,' he told me.

It goes without saying that Edgar Pask also tested the survival suit. He leapt into the frigid Atlantic off Shetland, the northernmost point of Britain. A fierce wind chill enhanced the test. The trial had to be terminated before the observers froze to death. Pask ribbed the others by complaining that he was 'too warm'.

There is little value in protecting the survivor from cold if he then dies of thirst. Water is the most important ingredient in our diet. Yet we are profligate with it; just breathing out wastes over half a litre per day. Hunger kills within weeks, thirst within ten days.

Mellanby's conscientious objectors determined that one had to drink a litre a day to remain hydrated when eating 'lifeboat rations' – the driest of dry foods. Small amounts of water made little difference physiologically but gave a psychological boost. The emergency water supply on lifeboats was pitifully small. The landlubber who designed the ration packs and water bottles clearly planned for only a day or two at sea, but being a castaway is an open-ended enterprise. It is impossible to allocate meagre resources properly when there is no way of knowing how long you will be adrift.

There can be few more aggravating ways to die than of thirst while floating on top of the largest reservoir of water in the world. The problem of drinking seawater is that the body can only get rid of the excessive salt by using greater amounts of water, thus dehydrating the body further. The Toronto team did trials on volunteers who ate only the emergency rations (800 calories a day) provided in a dinghy. Half of them eked out the freshwater supply by drinking 450 millilitres a day diluted with 280 millilitres of seawater. At the end of the experiment those who drank the seawater supplement had lost twenty-five per cent less weight than those on the fresh water alone. Tests revealed that they eliminated all the extra salt ingested and there were no deleterious effects on the blood or urine. Unfortunately there is no indication of how long the trial lasted. The results were locked away in confidential reports. Had they been published they might had saved Alain Bombard a great deal of discomfort.

In 1951 a young French doctor was called to treat the crew from a ship that had been wrecked on the coast. Dr Bombard failed to revive any of the forty-three victims. It alerted him to the horror of shipwreck. He read that every year 50,000 people survived the sinking only to perish in lifeboats, many of them from thirst. He decided to do something about it.

There were stories of castaways who drank seawater and survived. Teehu Makimare had only nine litres of fresh water to sustain him and his six companions (later four because two drowned) for sixty-four days. For almost half the time they drank seawater. Thor Heyerdahl diluted his freshwater supply with between thirty and forty per cent of seawater on the epic voyage of the *Kon-Tiki* raft across the Pacific.

The Board of Trade Merchant Shipping Notice was unequivocal:

> A belief has arisen that it is possible to replace or supplement fresh water by drinking seawater in small amounts. This belief is wrong and DANGEROUS.
>
> Drinking untreated seawater does a thirsty man no good at all. It will lead to increased dehydration and thirst and may kill him.

According to Bombard, the secret was not to wait until you were dying of thirst but to take small amounts of seawater from the beginning. Since the body *needs* sodium and acquires it mostly from sodium chloride, Bombard reasoned that providing one drank no more seawater than necessary to fulfil our required daily salt intake, no harm should result.

There was no doubt that drinking large amounts of seawater caused potentially fatal inflammation of the kidneys. What Bombard advocated was sipping seawater at first to sustain castaways until they organised a source of potable fresh water. Where would this water come from? Bombard noted that the water content of fish varied from sixty to eighty per cent. Surely they could be tapped like a rubber tree, but for water. We are often told that we should drink several litres of water every day in order to stay fully hydrated. This claim is based on research that calculated the necessary amount of water *in our diet*. Much of our moisture comes from the food we eat, not just from drinking.

Bombard planned a bold self-experiment. He would set himself adrift without food or water and attempt to survive exclusively on what the sea provided. His 'lifeboat' was a second-hand inflatable dinghy. It was four and a half metres long with a U-shaped pontoon closed at the stern with a wooden board. The central well was less than a metre wide. There was no motor, only a small sail. He called it *L'Hérétique*, to reflect his views on survival at sea.

A preliminary 'cruise' in the Mediterranean was planned to try out his ideas. As news of his adventure spread he was inundated by letters from people wanting to join him. One correspondent admitted to two unsuccessful suicide attempts but felt sure that Bombard had come up with a sure-fire method. Another volunteer, in the true spirit of self-experimentation, offered to let himself be eaten if things went badly. An Englishman called Jack Palmer seemed a better prospect. He was an experienced yachtsman and knew how to navigate. Bombard took an instant liking to Palmer and the small dinghy now had a crew of two.

They set off from Monaco on 25th May 1953 and sailed westward towards the Balearic Islands. A marine expert told

Bombard's pregnant wife that she would never see him again. Without fresh water both he and Palmer drank a couple of mouthfuls of seawater eight or nine times a day for ten of the following fourteen days without ill effects.

They hooked their first fish on the second day and squeezed out its 'juice'. At first it made them heave, but they soon got used to it and it quenched their thirst. In the following days their fish hooks came up empty and the cramps from hunger became 'almost more than we could bear'. It didn't augur well for long-term survival at sea.

After a landfall in Minorca, *L'Hérétique* was towed back out to sea to continue the journey. The combination of a freak wave and the towline capsized the boat. The mast and rudder were broken and the oars, radio, cameras, binoculars and sleeping bags were lost. They would all have to be replaced, but not until the craft made it into the Atlantic. They passed through the Straits of Gibraltar at night, unable to sleep as huge freighters threatened to run them down.

Bombard developed an abscess in his mouth. It became so painful that he had to lance it with a knife. There were antibiotics on board, but he refrained from taking them because an accidental castaway would not have access to such medication.

A short stay in Tangier allowed them to replace some of the lost equipment. Already Bombard was having reservations about *L'Hérétique*. It was three years old and showing signs of wear. A replacement failed to materialise. An adviser predicted that the rubber dinghy wouldn't survive more than ten days in the Atlantic. Palmer thought it was suicide to try and argued that they should return to the Mediterranean to complete the experiment. Bombard left a note for him: 'I am taking the responsibility of leaving alone . . . If I fail, then it will be the fault of a non-specialist.' Bombard was indeed no specialist. He was neither a mariner nor did he have the slightest idea how to navigate. Indeed, he was in the same boat as a typical castaway.

He did, however, have a watch, a sextant and an incomprehensible book on navigation so he set about taking readings of the angle of the sun. It didn't seem too difficult to fix his position. Unfortunately, it was more difficult than he knew. The rubber dinghy, with its toy sail, was unsteerable. From now on it would be driven by the wind and guided by currents to an unknown destination. Bombard had abandoned himself to the relentless ocean.

From the first day of his Atlantic voyage the currents carried *L'Hérétique* towards the Canary Islands, 1,400 kilometres to the south. Should it miss them, the next landfall was well over 6,000 kilometres beyond, on the other side of the Atlantic. The fish were biting and a bream even volunteered by leaping into the boat. Eleven days later the Canaries rose out of the ocean.

From Las Palmas Bombard was able to tell his wife that he had arrived safely. She told him that their baby daughter Nathalie had arrived too. He also acquired a radio receiver.

It couldn't transmit so there was no question of sending an SOS should he get into difficulties.

From Las Palmas he drifted westward on the Northern Equatorial Current, as if riding a river in the sea. The currents in the North Atlantic form a vast gyre over 8,000 kilometres in circumference. Somehow he had to stay in the main stream. If *L'Hérétique* drifted too far south he risked the fierce trade winds that whipped up storms at the slightest excuse. Should it veer to the north, he might become becalmed for ever among the seaweed rafts of the Sargasso Sea.

Fishermen had warned Bombard that fish were rare in the open ocean. His catches were barely enough to supply his need for water, so he took to sipping seawater. It tasted less salty than the Mediterranean.

He found himself in an ocean of storms. His tiny craft was dwarfed by giant waves. It surfed the crests and then slid down into the deep, dark valleys until he was surrounded by walls of water, imprisoned by the sea.

The dinghy was swamped and Bombard was as submerged inside it as he would have been in the sea beyond. Bailing was futile. He strapped himself to the mast and waited for the storm to abate. In another gale two days later the sail was torn in two by the wind. He replaced it with the reserve sail, which promptly blew away and was lost. He had to repair the original as best he could.

For weeks he baked by day and froze by night. Swampings

became so frequent that for the rest of the voyage his sleeping bag was a wet sack. He rarely slept at night for fear of the great waves capsizing his tiny dinghy.

In the sun every surface became frosted with salt crystals that absorbed moisture, keeping everything damp. The salt irritated Bombard's cuts and abrasions and by sitting all day he developed bedsores. His only relief was a small cushion. To his dismay he saw that it had fallen overboard and was bobbing in the waves a hundred metres away. He slung the sea anchor over the side. It was a drogue like a parachute that opened under water, slowing the boat's drift. He soon made it to the cushion, but turned to discover that the dinghy was drifting away at speed. The drogue must have snarled on its line and failed to open properly. Bombard swam towards the boat as fast as he could. He was a strong swimmer though not now at his fittest. But the dinghy was drifting too fast for him to catch. What a fool he had been. He was going to drown for the sake of a cushion. Suddenly the boat slowed. The drogue must have opened. In minutes he was relieved to be back on board.

Bombard caught a regular supply of fish. They provided him with water, protein and fish bait. Some bones served as fish hooks. Even so, he was rapidly losing weight. Many diets rely on recommending such a limited variety of food that the slimmer eats less and less because of the boring monotony of their diet. Raw fish every day would certainly do the trick for me. When a bird became snagged in his fishing line, Bombard anticipated savouring its chicken-like flesh. It tasted of fish.

By now he had been at sea for over forty days and should have been showing symptoms of scurvy. He was thousands of kilometres from the nearest lemon but he had devised a novel antidote. Bombard knew that whales, like humans, cannot manufacture vitamin C, but they don't suffer from scurvy. The vitamin must come from their diet and some whales feed exclusively on the tiny crustaceans in plankton. So every day he scooped up plankton with a fine-mesh net and swallowed two teaspoonfuls of the unappetising slurry. It worked.

Tethered to the dinghy by a line, he inspected the submerged rubber parts. To his horror he found that the repair patches that had been put on in Las Palmas had come unglued and were flapping loose. He became obsessed with checking for leaks. Every day he ran his hands over all the surfaces of the boat, feeling for wear. He put his ear to the pontoons like a doctor sounding the chest of a sickly patient. The slightest hiss would indicate an air leak.

Frequent storms added to Bombard's stress. A dead-calm sea could become angry in minutes. A gale that lasted for ten hours snapped the rudder arm. The tiny dinghy was thrown about like a kite in a reckless wind. Bullets of rain punched holes in the surface of the sea. He collected the fresh water in a small tarpaulin between his knees. The tarpaulin was so encrusted with salt that it tainted the water, making it saltier than the ocean.

Every day Bombard took his blood pressure and pulse, tested his strength and measured his urine. None gave reason for alarm, although he didn't realise that he was seriously anaemic. His emaciated body was covered in painful pustules. Sheets of skin peeled from his feet and his toenails became deciduous. There were pockets of pus beneath his fingernails that he lanced without an anaesthetic. Again he shunned medication because it was not in the spirit of the experiment. All he could do was to endure.

The lone castaway faces an enemy even greater than physical deterioration. It's called despair. Loneliness and the continual battle against the elements erodes morale. The melancholy Bombard began to talk to inanimate objects and accused them of plotting against him. His thoughts dwelt on his wife and the new child that he might never see. Every day he predicted that he would sight land, but only sea rolled over the horizon. Incessantly he checked the boat's position and his estimates of how far he had travelled. The figures didn't add up. He had no idea where he was.

The dinghy was never steady and it's not easy to take accurate readings with a sextant while on a bouncing trampoline. In lumpy seas it's not even possible to be sure whether you are sighting on the horizon or the top of a wave. From the outset he had made an error in his calculations. Although he didn't know it, he was ten degrees further east than he thought.

For every two days driven by the wind Bombard spent

ten days hardly drifting at all. By his fifty-second day at sea he had been becalmed for eighteen consecutive days. He was exhausted and was suffering from diarrhoea and haemorrhages. At night, flying fish fleeing from predators landed in the boat. It didn't matter. He couldn't face another rawfish supper or a swig of seawater. He began to think of being dead on arrival. Rain fell all around, but not on the boat. Even the gods were against him.

The very next day the first ship he had seen since leaving Africa stopped and the captain invited Bombard on board. He was served a snack and after an hour and a half of human contact he returned to *L'Hérétique* to complete his voyage. At last he knew his true position. He was still 966 kilometres from the nearest landfall.

His morale was restored, but the dinghy was again becalmed. How could he be stationary when the *Castaway's Handbook* assured him that the trade wind was at its strongest and most regular at this time of year? Then a storm arrived to repeatedly swamp the boat while he frantically bailed with his hat and shoe. When the weather abated, the boat began to leak.

On the sixty-fifth day he saw the beam of a lighthouse. The next morning he landed on a beach in Barbados. The experiment was over.

Bombard had lost twenty-five kilos (fifty-five pounds). A fish-and-plankton diet is seriously short of carbohydrate. His red cell count had dropped by fifty per cent. He had an all-body rash and his vision was temporarily impaired.

He had survived, although he had drunk nothing but fish juice for forty-three days and nothing but seawater for fourteen days. He had missed a trick by not biting into fish eyes, which a later castaway described as 'nuggets of pure fluid'.

Despite all Bombard's privations the current medical advice is still that the castaway drinks seawater at his peril.

Bombard all at sea in L'Hérétique.

Carnivorous and Coming This Way

'Into this dangerous world I leapt' – William Blake

One of Bombard's greatest fears was an attack by swordfish or sharks. In a contest between either of them and a rubber dinghy the winner would not be in doubt. He lashed a knife to the oar to defend himself and had to use it several times. One shark repeatedly buffeted the boat and lashed its tail as if taunting him. A shark's skin is covered in tiny sharp teeth. You can sand wood with shark hide. After one had scraped against the inflatable's bottom, it sprang a leak. From then on nothing would persuade Bombard to get into the water.

In 1987 I was working in the Philippines and considered taking a ferry between the islands. The locals advised against this. The boat would be very overcrowded and uncomfortable. They didn't exaggerate. The *Dona Paz* was built to carry 608 passengers, but the company had the official capacity raised to 1,500. She set sail with an estimated 3,000 to 4,000 people on board. She collided with a tanker and there were only twenty-five survivors. Three hundred bodies were recovered from the sea. All had been mutilated by sharks. For weeks afterwards fishermen found body parts in the stomachs of tiger sharks. Whether they ate living or drowned passengers is uncertain. I am glad I wasn't there to find out.

Bombard drifted from the North Atlantic into tropical waters and man-eating sharks are denizens of the tropics. Aren't they? An American millionaire once offered a $500 reward for proof that any live person had been attacked by a shark in temperate waters. There were no takers. In 1916 the director of the American Museum of Natural History in New York declared that there was 'practically no danger of any attack from a shark about our coasts'. Perhaps the sharks took this as a challenge.

The New Jersey shore was a paradise for bathers. The summer of 1916 was a sizzler and Charles Vansant waited until the early evening before running into the cool surf. He swam out beyond the other bathers, enjoying being alone in the sea. Then he turned back towards the shore. In water only a metre or so deep a great white shark clamped down on his leg. The blood in the water alerted a brave man to come to Vansant's rescue. There was a terrible tug-of-war. The shark refused to let go and was rising out of the water. Then it relented before it was beached.

Three doctors rushed to Vansant's aid – one was his father. The leg was all but severed from the body and blood was pouring out onto the sand. He became the first man in America to have 'Bitten by a shark' on his death certificate. A scandalous divorce hit the headlines next day and Vansant's death was relegated to a brief mention on the back page.

Five days later Charles Bruder, who worked at a hotel just up the coast, took his daily dip in the sea. He had heard of Vansant's death but agreed with the majority of people who expressed 'doubt as to the veracity of the story'. Bruder boasted of swimming with sharks in California and coming to no harm. He was reputed to be the strongest swimmer on the beach and he swam out further than all the rest. But he was not alone. There was a massive explosion in the water and a woman on the beach shouted, 'The man in the red canoe is upset!' There was no canoe. It was Bruder's blood. Lifeguards in a skiff arrived promptly at the scene to see Bruder flung cartwheeling into the air. The shark struck repeatedly, shaking him 'like a terrier with a rat'.

With Bruder's horrific death even the newspapers now admitted there was a monster on the loose. And it was heading north. Coney Island and New York were only fifteen hours away at the shark's most leisurely cruising rate.

Asbury Park had guaranteed the 'absolute safety of its bathers'. But after the boat containing the captain of the lifeguards was attacked by a large shark 'a great many bathers are rather scarce'. It was not shaping up to be the most prosperous summer for the seaside resorts of the New Jersey coast. Their mayors issued a joint statement complaining that trade was being hurt 'without reason'. The Director of the Bureau of Fisheries told the public 'not to be unduly alarmed or deterred from going bathing'. Although no one saw *him* splashing about in the surf.

That same day a big shark was spotted in New York Bay. A policeman emptied his revolver into it. It just swam away. Opposite Staten Island is a small estuary called Matawan Creek. One of the boys larking about there felt something big scrape past him in the turbid water. He scrambled out and found that his chest was grazed and bleeding. The next morning a retired sea captain out for a stroll saw a huge shark rushing upstream towards the town. He ran as fast as his old legs would carry him to warn the townsfolk. They just laughed, thinking it was another of his salty tales.

That afternoon the boys were back swimming in the creek, 'bombing' into the water and making as many splashes as possible. It was irresistible to the shark. One boy suddenly vanished, only to reappear screaming and in a shark's mouth. Then he vanished for ever.

Two men volunteered to go duck diving to search for the body, no easy task in the murky water. One of the men, Stanley Fisher, surfaced and shouted 'I've got it.' Instantly the water around him began to boil and he screamed, 'He's got me.' As he tried to drag the boy's body towards the bank he was hauled down time after time. The shark snatched back the boy and Stanley was dragged up onto the bank. Half his thigh was gone and the flesh around the wound resembled bloody rags. He muttered that he had seen the shark below feeding on the boy and as he'd snatched the corpse away the shark had turned on him. The nearest hospital was over two hours away. Stanley was conscious throughout the journey, only to die on the operating table.

Meanwhile, downstream another bunch of lads were bathing, unaware of the drama. The retreating shark grabbed one of them, but a plucky fellow dived in and snatched him from the shark's jaws. As the boy was being hauled out of the water the shark took a final bite out of his leg. Below the knee his left leg was just ribbons of flesh, yet with prompt medical help he survived.

Two days later a boat trawling for fish was dragged backwards. When they hauled up the net a large shark's head appeared over the stern, snapping at them. The fishermen beat it to death with a oar. When the shark's stomach was opened, out tumbled the rib of an adult human and the shin bones of a child. Perhaps this was the killer shark. It was impossible to say whether all the mayhem was caused by a single shark or by several.

A board member of the Zoological Society searched for sharks around Long Island. In 1916 alone he caught 277 and shot over a hundred of them. None matched the description of the killer. Nonetheless sharks were common in these cold waters. Many years later, the seaward side of Long Island was found to be a hotspot for juvenile white sharks (popularly known as great white sharks). In 1964 a great white was caught off Long Island that was 5.25 metres long and weighed almost two tonnes.

In 1916, although witnesses had provided detailed accounts of five shark attacks and four victims had died in a period of only two weeks, the scientists were still in denial. A zoologist insisted that there was no reliable record of an unprovoked shark attack. The experts of the day worked in museums and laboratories and studied the anatomy and classification of fishes. They examined specimens small enough to fit into their pickling jars. Every zoology student dissected the dogfish a dozen times. They could tease out the blood vessels or the cranial nerves, but were taught nothing about the private lives of sharks. Indeed, little was known of their behaviour and ecology.

The first person to study sharks beneath the ocean was an Austrian called Hans Hass. In 1939 he and his student friends organised a diving trip to the Caribbean. They took a bucket diving helmet, like the one that Kitching used, swimming goggles and home-made flippers. They had several cameras to capture images of fish and especially sharks in their natural environment. They took 4,000 photographs, some of them in colour.

Now that diving is commonplace it is difficult to appreciate the risks they were taking. The Caribbean was teeming with sharks that had never met a human under water. How would they react? Which ones were benign and which were dangerous? There was a real possibility that the first dives of the expeditioneers would be their last.

During a dive by Hass and his friend Jörg, three sharks appeared and torpedoed straight towards them. It was so frightening that Jörg gave out a piercing yell. All three sharks turned and fled. So yelling became the divers' defence when threatened. When Hass tried it in the Mediterranean and the mid-Atlantic it had no effect whatsoever.

Hass read zoology at the University of Vienna and later received his doctorate from Friedrich-Wilhelm University in Berlin. He spent his spare time trying to raise money for another expedition. The first book of his underwater adventures was published in 1939 and a leading magazine serialised his account of the Caribbean trip. He also sold prints of his underwater photographs and went on lecture tours.

On most of his dives Hass held his breath, as did the reader. What he needed was a breathing device that would allow him to retain the freedom of the skin-diver. His requirements were put to Drager, the company that made the escape equipment for use in U-boats. They provided him with modified oxygen re-breathers in which pure oxygen was supplied from a small cylinder into a bag like a life vest from which the diver breathed. Caustic soda in the bag 'scrubbed out' the carbon dioxide from the exhaled air and allowed the diver to use the remaining oxygen. The gear gave substantial time underwater, but was dangerous. As Jack Haldane had confirmed, pure oxygen is potentially poisonous under pressure. In 1942 Hass became the first to use self-contained diving equipment for research. The aqualung was still just a glimmer in Cousteau's goggles.

There were accusations that Hass's shark photographs were fakes, which made him determined to get even better ones with divers and sharks in the same shot so there could be no doubt. His next expedition was marked by misadventure. Their ship caught fire before leaving port. When eventually they set sail it began to sink. So they joined forces with dynamite fishermen to photograph the behaviour of sharks that were attracted to the dead fish.

When the war ended Hass set off to study sharks in the Red Sea. On arrival at Port Sudan he was immediately told of a ship's passenger who had fallen overboard and been torn to pieces by sharks under the gaze of his fellow travellers.

The friendly British Commissioner recommended a site where 'the water was so full of sharks that you could put an oar into the water and it would remain standing upright'. Hass was lowered into the murky sea armed only with a length of picture framing whittled to take a harpoon point. He met sharks aplenty.

Even on land it wasn't safe. Torrential storms swept through the town and some residents were washed down to the sea to have their heads bitten off by sharks. Worst of all, according to the Commissioner, the golf course was flooded.

On his return to Austria, Hass found that the public were just as fascinated with sharks as he was. At a lecture attended by the Minister of Education he revealed that his field trip had been supported by donations from classes of local schoolgirls and a masked all-in wrestler. The Minister was sufficiently embarrassed to sponsor his next expedition to the Red Sea.

Hass's beautiful secretary Lotte became his diving companion. On their second dive she found herself alone underwater. A shark appeared and patrolled back and forth in front of her, assessing her first with its cold right eye, then with its icy left. She was terrified, but it departed. When Hass appeared she tried to tell him of her narrow escape, blabbering through her mouthpiece. All he did was to confirm that there was something wrong with the camera.

Sharks needn't be big to be dangerous. Hass caught a young shark by the tail. It was supple enough to whip round and grab his arm in its mouth. It then swam off dragging him with it. By the time it let go Hass was bleeding profusely. The flesh was hanging from his wrist in shreds. The starter cord for the outboard served as a tourniquet and he was rushed to hospital. Hass was out of the water for three weeks before he lost patience and removed his own stitches.

He had felt the chomp of a shark, but the critics' teeth were sharper still. They suggested that the sharks must have been stunned before being approached. Worse still, the value of pictures taken of animals acting normally in their natural environment was not appreciated by a well-known biologist: 'Any good aquarium picture is of greater scientific and instructional value than similar shots taken in nature, even at the risk of one's life.'

Hass had a theory that sharks homed in on the distress noises of fish. To enable him to study this, Philips provided some sound gear on which he recorded the flutter of harpooned fish. When played back underwater it attracted sharks from a distance. Later other workers recorded artificial low-frequency sounds and found they were also attractive to sharks, but only if they were pulsed. Continuous sounds had no effect.

Hans and Lotte married and honeymooned on the Great Barrier Reef. On their arrival the local doctor informed them that 'Only last week a young couple were eaten in the harbour.' He assessed their life expectancy as unlikely to exceed a fortnight.

While in the Maldives with fellow zoologist Irenäus Eibl-Eibesfeldt, Hass filmed the shark's ability to find food that it couldn't see. Eibl-Eibesfeldt harpooned a grouper and then hid the corpse in a hole on the reef. Sharks rapidly appeared and snuffled around trying to locate the bait. The first to find it chomped it in half. This aroused the others to start snapping at the successful shark and at Eibl-Eibesfeldt who was sitting perfectly still while Hass filmed the proceedings. Eibl-Eibesfeldt assumed that he was of more interest because he had handled the bait. The experiment was repeated many times with the same result.

Nowadays, when every schoolboy can recognise a manta ray, it is difficult to imagine the excitement the public felt when they first saw Hass's pictures. Every creature he photographed was strange and huge and potentially dangerous. The almost naked divers weaving between the predators seemed small and vulnerable. Hass never filmed even the most voracious sharks from within the safety of a cage, in case it induced unnatural behaviour.

The Second World War saw numerous pilots ditching into the sea and many thousands of sailors abandoning sinking ships. The survivors often told of being harassed by sharks. It happened far more frequently than we know, for those most troubled by sharks probably failed to file a report.

The US Navy grasped the magnitude of the problem in 1945. In July of that year the heavy cruiser USS *Indianapolis* delivered a vital component of the atomic bomb to the US airbase on the Pacific island of Tinian. From there the *Enola Gay* would fly to Hiroshima.

The *Indianapolis* departed for the Philippines to resume its role as flagship to the US Fifth Fleet. Headquarters failed to warn its captain that a Japanese submarine was active on its designated route. The cruiser was unescorted and had no sonar. It was, therefore, blind to submarines.

The Japanese submarine *I-58* had excellent sonar and could hardly fail to register such a huge ship. At midnight on 30 July the submarine fired six torpedoes at the *Indianapolis*. Should they miss, the sub also had a giant fourteen-metre-long torpedo strapped to its belly with a kamikaze pilot on board. It was packed with high explosives and could slice a ship in two. It wasn't needed. Two of the torpedoes destroyed the entire bow section of the *Indianapolis* and she sank within fifteen minutes.

Of the 1,196 men aboard, between eight and nine hundred survived the sinking, although fifty soon succumbed from their injuries. Those survivors in the worst state were put on twelve small life rafts with canvas pontoons stuffed with kapok. They had to sit on the pontoons as the rafts had no floor, just a net of ropes to brace the feet against. The majority bobbed in the waves, some with life jackets, some without. They clung to ropes to stay together. There were two large groups, one of around four hundred men, the other of 150, plus many smaller groups. Every group thought they were the only survivors.

Most of the rations in the rafts were soaked and useless. Many of the water containers were empty or contaminated with seawater. The officers tried to ration the supplies. Every man was allocated one cracker, one malted-milk tablet and a taste of water per day. From the start the men were drinking seawater, either deliberately or because it was impossible not to swallow some water and oil in the rough sea.

Sharks arrived on the first day and were beaten off with the rafts' paddles. On the second day they arrived in numbers to feed on the dead bodies. Some men fished, but everything they caught was stolen by the sharks. So they gave up. Then the sharks turned against the men. When a shark singled out its victim, neither yelling nor beating of the water would deflect it. There would be a scream and the water turned crimson as the shark butchered its victim.

By the third day men were becoming delirious and delusional. Some swam away towards imaginary islands. Others became violent and fought over life jackets, although as men died the supply became ample. Perhaps as many as twenty-five were killed by their shipmates.

To add to the men's despair the kapok in the jackets was gradually becoming waterlogged. They were sinking lower and lower in the water, and the sharks kept coming back. One man turned to his companion whose head was dipping towards the water. He just flipped over to reveal that the entire lower half of his body was missing. A shark had chopped him in half.

When help eventually came on the fifth day the rescuers had to use gunfire to keep the sharks at bay. Every body they pulled from the water had bits missing. Half were stripped to the bone. About five hundred men had died in the sea. Most had been eaten by sharks, but there is no way of knowing how many were alive at the time. One survivor claimed that he had seen over eighty attacks on sailors.

Most of the 350 species of shark are 'chinless cowards'. Thirty-five species have been known to attack man, but only ten are frequent offenders. Yet polls reveal that one of our greatest fears is being attacked by a shark. We are not reassured that more people are killed by falling coconuts than are eaten by sharks. The remoteness of any risk of being

attacked is not the point. It is the fear of being systematically dismembered that stokes such terror. It rekindles the
ancient fear of our ancestors for whom death by wild beast
was common. We have tamed or shot almost all the big
terrestrial carnivores. Today vehicles are our greatest pred-
ator, yet we accept the carnage on the roads as a risk of daily
life. Being eaten alive invokes a much higher order of dread.

After the *Indianapolis* disaster a perceptive navy researcher
announced that fear of death and dismemberment by sharks
was a major morale problem among ditched flyers and
survivors of sunken ships. The US Navy began to fund the
search for an effective shark repellent.

The remit was to find a chemical that could be leaked into
the water to form a protective cloud around a floating man.
It had to be sufficiently concentrated to repel a shark, yet
innocuous to the man. They tried hypochlorites, chemical
warfare agents and cyanide — having forgotten the 'innocuous
to man' requirement. No compound met the criteria. In any
case, how would you maintain a protective cloud in an ocean
of diluting water beset by turbulence and currents?

Some sharks had been observed to avoid decaying fish. Of
all the chemicals given off by rotting carcasses someone
decided that ammonium acetate was the repellent substance.
So they added a toxic metal and came up with copper acetate
mixed with a dye to obscure the shark's vision. Even though
preliminary trials indicated that the dye was avoided more
than the repellent, copper acetate was supplied with all life

jackets and life rafts. The packets were labelled 'Shark Chaser'. Although they provided great psychological comfort to the troops it's unlikely that they chased away many sharks. Eventually the US Navy admitted that 'Shark Chaser' was perhaps not the ultimate deterrent.

It was decades later that a diving zoologist called Eugenie Clark observed that sharks sampling the Moses sole spat it out in disgust. The fish was found to secrete a surfactant (detergent) repellent to sharks. Donald Nelson and Wesley Strong developed an air-powered 'syringe gun' to deliver a dose of the detergent into the mouth of an advancing shark. Its sea trial took place in 1991 at the aptly named Dangerous Reef in South Australia. It is an assembly place for white sharks. The original idea was to squirt the sharks from the safety of a stout cage, but the wily fish refused to come close enough.

A more hazardous method was tried. The gunman had to be close to the shark's mouth so Strong crouched on the swim deck at the stern of the boat. It is a shelf only thirty centimetres or so above the sea's surface. A shark was lured in by trailing bait behind the boat and then drawing it up until it was just out of the water. As the shark lunged open-jawed with its head in the air, Strong fired a dose of detergent into its mouth. Every shark responded vigorously to being dosed, making a rapid retreat and not returning for days, if at all. A photograph makes it clear that Strong was dangerously close to the shark's mouth. The slightest misjudgement would indeed have provided a stern test for the deterrent.

His fellow researcher Donald Nelson was also not averse to risk. On an earlier project a shark had charged him. His dive buddy photographed him fending off the beast with his hand. Its teeth were within centimetres of his groin. The assailant had to be shot to save Nelson's life.

After this experience Nelson studied shark behaviour from the comparative safety of a small one-man submersible shaped like a shark complete with a tail and fins. The submersible 'confronted' and pursued grey reef sharks (close relatives of great whites) thus provoking aggressive swimming displays and attacks. It was attacked fifty-seven times. Most often the sharks charged the submersible's transparent observation dome. Following their violent lunges the plexiglass (perspex) dome became deeply scarred and weakened. Had it given way, Nelson would have drowned.

His companion was Scott Johnson, who devised a novel way to be safe from sharks – maybe. He designed the 'Johnson Bag'. Imagine a large black plastic sack like a giant novelty condom. The open top had an inflatable yellow collar. The man climbed inside the sack and peeped nervously over the collar, looking out for sharks. The idea was that the bag kept the survivor's body fluids from leaking into the sea and alerting a shark to his presence. Also, as he relaxed with confidence in his condom, he was not splashing about, which would be another signal to sharks. Since Johnson invented the device he was the one to test its effectiveness in a pool full of sharks. He was investigated, probably because the yellow colour of the collar is highly attractive to sharks, but he was not attacked.

Several non-scientists also became obsessed with shark biology. David Webster knew danger. In the Second World War he had been a member of 'Easy Company', immortalised as the Band of Brothers. After the war he became a journalist with the *Saturday Evening Post* and the *Wall Street Journal*.

The quiet life didn't suit Webster and he turned to the sea for excitement. He surfed and dived and became interested in sharks. They represented all that was mysterious and dangerous about the ocean. He studied them underwater by swimming among them. In September 1961 he left Santa Monica pier in California in search of sharks. Perhaps he found them for he never returned. His boat was found drifting eight kilometres offshore. Its tiller was gone and so was Webster.

Webster's book on sharks was published posthumously the next year. He had collected an archive of stories. One tells of a sailor stranded on a life raft and being dogged by a shark. None of the suggestions in the on-board survival manual deterred the creature. In frustration he tore its pages into shreds and cast them onto the sea. The shark followed the paper trail and didn't return.

Another enthusiast, Michael Rutzen, skippered a dive boat and lowered tourists in cages to the sea floor off South Africa. He became obsessed with white sharks and felt he could only get to understand them if they met on the shark's terms.

He snorkelled and SCUBA-dived with them and learned

that they responded to his body position. If he curled into a ball it attracted their attention. When the sharks came too close he uncurled and they usually turned away. If he swam away, a shark would follow. As they read his posture so he read theirs. A gaping mouth was a sure sign of aggression. Really? I should have guessed.

Rutzen usually conferred with sharks in relatively shallow water to minimise the risk of attack from below. When there were several great whites in the vicinity he backed onto the reef to protect his rear. He was particularly cautious when they were feeding, in case he was mistaken for the side order. If the sharks got too frisky he didn't try to surface. Instead he went below, which he considers to be the 'power position'. Despite his astute reading of the sharks' body language, his body is covered with scars donated by sharks that clearly weren't paying attention to his.

Theo Ferreira, another South African, developed an interest in sharks that his son Craig inherited. Together they developed the White Shark Research Institute on the Cape coast. Their main aim is to save the white shark from extinction, for vastly more sharks are killed by man than the other way round. The jaws of a great white make a fine trophy, or so they say.

The Institute is not far from 'Shark Alley' where white sharks still congregate in appreciable numbers. They are attracted to the boat in the traditional way by ladling 'chum' into the water. Chum is an irresistible bouillabaisse of sardine mush, fish guts and blood. Every shark that comes close enough is numbered by jabbing it behind the dorsal fin to fix a tag. This means that the individual can be identified when seen again, and the ratio of tagged to untagged sharks seen subsequently gives an indication of the population size.

Researchers also sample the shark's blood. A syringe is inserted by hand to prevent contamination by seawater. Craig Ferreira lies on the swim deck of the boat with the shark immediately below him. He has lost count of the number of times he has had to jump out of the way of the shark's teeth.

To determine whether a male is sexually active, Craig has to feel the external 'claspers' that serve as a penis. Sharks violently object to such foreplay and should they bite him it would not count as an unprovoked attack.

Many of the Ferreiras' underwater observations of shark behaviour were made from inside a cage. The cage is not a close-barred steel prison immune from all attack. It resembles a large circular basket of steel mesh that looks little stronger than chicken wire. It wasn't designed to resist a charging shark, merely to deflect it. The researchers inside must feel it's like a game of dangerous dodgems. More worrying still, the top half has only occasional metal poles with large openings between them. These 'windows' give the researchers an unobstructed view of the advancing sharks and are easily big enough to allow the head of a great white to enter. Indeed, one agitated shark got *into* the cage to hassle two cameramen. They described the experience as being locked in a coffin with an excited chainsaw. Several times researchers have been trapped in the cage when five-metre sharks have grabbed the bars with their teeth and have shaken the cage like a toy. When sharks once became entangled in the cage's cable, the entire cage and its human contents were almost dragged down to the bottom.

Craig Ferreira has had his air hose severed six times and he has often been no safer on the surface. Angry sharks almost scuppered the boat on more than one occasion and frequently he was almost hauled into the sea. Ferreira admits that they have had lots of close calls. He knows that a white shark can end it for you whenever it wants. In his words, 'It's all fun and games until someone gets munched.'

I have personal evidence that marine biologists view sharks with both wonder and apprehension. I was sitting in a boat when someone spotted a large fin breaking the surface and shouted 'Shark!' All the divers in the boat leapt into the water and all those in the sea scrambled back into the boat.

That certain smile, circa 1700.

Into the Abyss

'The abyss below becomes a pleasant walk. A little further — why not. And then suddenly comes the end, without one even being aware of it' — Hans Hass

Much of my early career was spent underwater. Not in search of sharks, but studying the ecology of coastal communities. The deepest I ever needed to go was fifty metres. Deep down in an Irish sea lough the water was an eerie luminescent green, as if lit for the entrance of the Demon King. And it was absolutely calm. I was used to being jostled by waves and found the stillness unnerving.

For eight thousand years silt had washed in from the surrounding land and accumulated. The bottom sediment was already twenty-one metres thick, some of it so soft that I just sank in as if there were nothing there. As the black cloud enveloped me and closed over my head, I tried not to panic. But which way was up? Which way was out?

Nothing much happened down here, except for death. By autumn, the bottom dwellers had used up all the oxygen and everything died. Tiny tubular worm coffins bristled from the mud and abandoned burrows gaped like mouths gasping for air. I was suddenly aware of the air escaping from my regulator and realised that I was the only living thing in the landscape. So why did I have this feeling that someone – or something – was following me?

Above me trembling medusae of air rose, expanding towards the surface. I became aware that there was more beauty in the rippling rainbows of liberated bubbles than in all the paintings in all the art galleries of the world.

Perhaps I was falling under the influence of nitrogen narcosis, an inebriation caused by breathing nitrogen under pressure. A diver using compressed air has no choice. Seventy-eight per cent of air is nitrogen. It causes another problem: decompression illness. Underwater the pressure causes greater amounts of the air that a diver breathes to dissolve in the body's tissues. On the return to the surface the pressure is relieved and the diver must allow sufficient time for the excess nitrogen to be respired away safely. If he ascends too rapidly, the gas comes out of solution and the blood may fizz like deadly champagne. Bubbles of nitrogen can lodge in the joints and block blood vessels. The result is the bends. Peter Throckmorton described the symptoms suffered by sponge divers: 'You can be paralysed in your sleep, so that you wake up to find yourself a cripple for life. It can choke you to death, kill you instantly, or twist you into a screaming lump of agony with awful pains in your joints. You might get off with only a headache or an itching rash.'

Nitrogen was the major obstacle to deep-diving operations. A young Swedish engineer thought he knew how to solve the problem. Arne Zetterström devised specialist diving equipment. His large water-jet device was used to excavate tunnels beneath the wreck of the seventeenth-century warship *Vasa*, once the pride of the Swedish navy. Steel hawsers were passed through the tunnels to form a lifting cradle. The vast ship is now displayed in a custom-built museum in Stockholm.

For his military service in 1943 Zetterström was drafted into the navy's diving branch. He puzzled over the problems of rescuing sailors trapped in a disabled submarine. How could a diver get down to the depths at which submarines operated?

Air was too dangerous so he would have to breathe something else. The nitrogen had to be replaced. Dense gases were no use because under pressure they became so viscous that breathing was difficult. Among the lighter gases only helium and hydrogen were suitable. Helium was expensive and unavailable in Sweden. So it had to be hydrogen, which Zetterström could make himself.

The drawback was that a combination of oxygen and hydrogen is explosive. However, Jack Haldane had shown that the mixture is safe if the oxygen doesn't exceed four per cent of the total. Oxygen constitutes twenty-one per cent of the air we breathe. How can a diver survive on only four per cent? At depth there is no problem because, although the ratio of the gases in the mix remains the same, they are 'concentrated' by the pressure so at only thirty metres down each breath provides the diver with four times more oxygen than at the surface.

Zetterström's plan was to breathe compressed air down to thirty metres, then switch to his mix of low oxygen with hydrogen. He realised that this could not be done with a simple changeover, since at the juncture where air and hydrogen combine the mixture becomes explosive. But 'by ventilating the air out with a mixture of four per cent oxygen and the rest of it *nitrogen*, the risk of explosion is completely eliminated'. So a brief sojourn on a low oxygen/high nitrogen supply enabled him to use up the excess oxygen in his lungs. It was an ingenious solution.

Zetterström was to be the experimental diver in a series of trials from a naval vessel to test his theory. On a freezing winter's day in the Baltic, despite rough seas and a snowstorm, he descended to 110 metres. On his return he suffered a mild bend in his arm and felt dizzy and nauseous for a couple of days. This did not discourage him from attempting a further dive to 160 metres.

On 7 August 1945 the ship was cluttered with senior naval officers, including Zetterström's father who was a commodore. With so many chiefs around it was difficult to know who was in charge of the operation. Zetterström was dismayed to note that the experienced ship's crew had been replaced by new recruits. He had reservations and an officer suggested they should call off the attempt. Apparently Zetterström's father wouldn't consider a cancellation.

Zetterström used the standard hard-hat diving suit although, as the oxygen/hydrogen mix was not just explosive but also inflammable, he also wore fire-resistant underwear made from fibreglass.

He climbed onto the diver's cradle in which he would be lowered over the side. The cradle was just a wooden platform held on stays. It carried large cylinders containing the three gas mixtures. Zetterström would be responsible for changing from one mixture to another during the descent and ascent and would instruct the boat party to take the cradle to the appropriate depths. During the earlier trials they had found that breathing hydrogen gave his voice such a high nasal pitch that the boat party had had difficulty understanding what he said. He replaced the telephone with a telegraph key. It was vital that the communication was clear to ensure that the change of gases took place at the correct depth.

All went well on the descent to 160 metres, double the world record. On the return he halted at the agreed fifty-metre mark to decompress and then change the gas mixture. Although the main winch had stopped, the cradle began to tilt and rise. An additional line had been attached to the cradle to keep it stable in the strong current and this line was still being pulled in. The cradle's platform tilted alarmingly and Zetterström was hanging on as best he could. The cradle continued to rise to almost ten metres. At this depth he was breathing far too little oxygen to keep him alive and was in no position to change his gas supply.

Those on the surface noticed that something was wrong and rescue divers were sent down. They secured him while the platform was lowered to sixty metres to restore the pressure and allow him to decompress slowly. It was too late. Zetterström died of asphyxiation and a severe bend caused by rapid decompression. He was only twenty-eight years old. His technique was proven, but with his death it was stillborn.

In 2004 a contingent of members from both the Swedish and British Historical Diving Societies visited Arne Zetterström's grave in the family plot near Nynäshamn. His headstone bears a diving helmet embraced by a wreath. They cleaned the stone and laid flowers in memory of a man who, in the words of a naval surgeon who worked with him, 'never hesitated in staking his own security in pursuing a task'.

A Royal Navy diver called George Wookey had descended to 180 metres in standard diving gear, breathing a mixture of oxygen and helium. This avoided nitrogen narcosis, but not decompression. He took only twelve minutes to descend but needed six hours and twenty-one minutes to return safely to the surface. Clearly this was another inconvenience of deep diving.

In the early 1960s two Austrians decided they could alleviate this problem. Hannes Keller, a mathematician and a keen sports diver, enlisted Albert Bühlmann, a lung physiologist at Zurich University, to collaborate on a deep-diving project. He believed that with the right mix of gases decompression times could be greatly reduced.

They had access to a large new computer at the university and used it to calculate the decompression times required for different gas mixtures. With little technical support and only an empty oil drum for a diving bell, they tested different mixtures beneath Swiss lakes. On one trial Keller used so many different gases that he had four cylinders on his back and four strapped to his chest. He eventually reached a depth of 229 metres and returned to the surface in an astonishing thirty-four minutes.

This attracted the interest of the US Navy and Shell Oil. With substantial funding and support, and their secret mixture of gases, the two Austrians were ready to shatter the world depth record. Bühlmann would be the team doctor on the surface and Keller would lead the diving party. He needed a fellow diver.

Peter Small was a journalist specialising in medicine and science. He was one of the founders of *New Scientist*, Britain's premier popular-science magazine. He was also a diver and co-founder of the British Sub-Aqua Club, which, thanks to his suggestion of forming regional branches, became the biggest diving club in the world.

Small was adventurous. Reputedly the youngest captain in the British Army, he joined Vivian Fuchs's polar party. He also paddled across the English Channel in a canoe to test a theory about the currents, and bobbed about in the River Thames for hours testing a survival suit that he had designed.

Small was a visionary who believed that diving should have a serious purpose. He had spent a couple of years as a commercial diver in the Persian Gulf inspecting oil rigs. In the first issue of the British Sub-Aqua Club's magazine in 1955 he wrote, 'The real excitement lies in the opening up of the undersea frontiers.' He looked forward to 'a prospect of exploration and science of staggering proportions'. If we could dive down to 300 metres the entire continental shelf and all its riches of oil and minerals would be readily accessible. No wonder he wanted to play his part. Keller warmed to Small's enthusiasm and quiet charm. After a diving audition he was on board.

Before Small set off to California to join the team a friend asked, 'What mix of gases will you be breathing?'

'No idea,' he replied. 'I leave that to Hannes.'

The friend thought he was foolish: who would risk his life without knowing what he was doing? How could he assess the level of risk?

Keller and Bühlmann kept their recipe of gases secret for years after the dive because it was a valuable commodity.

Clearly the nitrogen had been replaced with helium and pure oxygen was being used prior to the dive and in the final stages when the divers were back in the shallows. But no one knew for certain what was in the cylinders.

Peter Small married Mary Miles on 12 October 1962 and in less than two weeks they were in California aboard a ship off Catalina Island where the record-breaking attempt would take place. Some believed that Peter was having second thoughts, but Mary was thrilled with the heroic exploits of her new husband. He also had a lucrative contract from a magazine to write his personal account of the adventure. He couldn't back out now even if he'd wanted to.

Keller had commissioned a diving bell with a lockable hatch on the bottom so that it became a sealed chamber in which the divers would make the descent. The plan was that they would descend to 305 metres (over 1,000 feet) in the chamber, then pop outside for five minutes to plant the Austrian and American flags. There were closed-circuit TV cameras on the chamber to record the ceremony for the sponsors and the press.

The descent took place on December 3rd and went according to plan. They opened the hatch and dropped out onto the sea floor a metre or so below. Keller got tangled in the large flags and couldn't see a thing. It took him minutes to get free and plant the flags. Because of the extreme pressure at this depth their breathing apparatus only gave them four minutes' breathing time outside the chamber. They returned hastily through the open hatch. It was only the pressure of the gas inside the chamber that kept the water at bay.

What they should have done immediately was to have topped up their breathing apparatus from the cylinders of gas, but instead they struggled to close the hatch and purge the water out of the chamber. No sooner was this achieved than Keller collapsed. Small was in a daze. He was instructed by Bühlmann above to take off his mask as the oxygen supply in the apparatus must be very low. But he froze and eventually he too fell unconscious.

There was another problem. The chamber was losing pressure. It was raised to sixty metres and two divers were sent down to locate the leak. Dick Anderson was a very experienced deep diver who had been technical consultant on Disney's epic *20,000 Leagues Under the Sea*. The other diver was Chris Whittaker, a British student at UCLA who was hoping to go on to graduate studies in marine biology.

They couldn't find the leak and returned to the surface. Whittaker had a problem with his life jacket that brought him up too fast and he had a nosebleed. The pressure in the chamber was still dropping. This was serious for the occupants. Anderson decided to have another look at the chamber. Against advice, Whittaker insisted on going down with him. 'Peter is my friend,' he said, 'I must go.' His life jacket would not deflate so he slashed it with a knife. He would no longer have the security of a passive lift to the surface should anything go wrong.

This time Anderson spotted a small gap at the edge of the hatch. He heaved against it with his back and eventually it closed. But he couldn't find Whittaker. He had simply vanished. He was nineteen years old.

Inside the chamber both Keller and Small came round. The chamber couldn't be opened until the men inside had decompressed. This took four and a half hours. Small fell asleep again and never awoke. A prolonged shortage of oxygen had impaired his circulation so his body had been unable to eliminate nitrogen efficiently. He died of the bends.

There was an inquiry and neither the Sheriff's Office nor the Chief Medical Examiner was satisfied with Keller's testimony, which was at odds with that of Bühlmann. A committee of experts was set up to assess the evidence. They could not decide whether Keller was confused or was 'evading issues to protect his interests'. Within the team Keller's previous successes had fuelled a feeling of euphoria. They were sharing in the glory of a much-publicised record-breaking dive. Making the first 1,000-foot dive became a magical prize, like breaking the sound barrier.

Later Keller revised his account of what had happened and admitted to mistakes. Prior to the dive he had discovered that there was a leak in one of the tanks of gas in the chamber. It was only half full, which greatly reduced their safety margin should anything go wrong.

He confided to Hans Hass why he had decided to continue with the attempt and risk the flag-planting ceremony: 'This was the situation: barely enough gas in the equipment carried on the back. On the other hand, the team in top form, weather perfect. Personally, a strong fear that it might all be called off. Knowing that one never has perfect conditions ... I decided to make the attempt.' What began as a demonstration of a new technique for exploring the ocean became a record-breaking attempt to impress the sponsors and the public, with all the additional pressures that brings.

Mary Small had been married for just three weeks when she saw on the TV monitor her husband's collapse. Perhaps she also saw the photograph of Keller in a magazine. The caption beneath his defiant face read: 'My system made no mistakes!' She attended the meeting at which Keller admitted to making mistakes. A few days later she committed suicide.

Keller became a consultant to Shell, advising on deep-water operations. The oil companies had recently begun to use helium divers to service their rigs. The bends were a common occurrence. To test whether they had decompressed long enough, the divers hopped around the chamber. If they collapsed, they needed a little longer. They relied on the manufacturer supplying the right mix of oxygen and helium. When the gas man was on holiday, his replacement got it wrong and all the divers had hallucinations, seizures and the conviction that their feet were being electrocuted. Some saw haloes while driving home and could feel bubbles passing through their blood vessels.

Keller conducted simulated dives in pressure chambers down to the equivalent of 300 metres with progressively shorter decompression times. Helium/oxygen diving is now routine for commercial deep diving, but the emphasis has switched from speed of decompression to having greater safety margins. The other development has been the widespread use of diving computers that automatically calculate a diver's decompression schedule. The algorithms that constitute their brains were developed by Albert Bühlmann.

Putting a 'wet' diver 305 metres down into the sea was a remarkable feat, but Keller had merely scratched the surface. The average depth of the world's oceans is 4,000 metres. If Mount Everest were dropped into the deeps of the Pacific, it would not reach to within two and a half kilometres of the surface. No mysterious brew of gases will allow us to go there. Yet we *have* made the journey.

William Beebe was an American ornithologist who widened his interests to include fish, the birds of the sea. After hundreds of dives in the shallows, his gaze turned longingly to the green depths far beyond the reach of his diving helmet.

His plans for a deep dive were published in the *New York Times* in 1926. An engineer and diver called Otis Barton wrote to him with a design for an underwater chamber. Barton, with his own money, built a large metal sphere with a manhole at one side and fused-quartz portholes on the other.

His bathysphere ('deep sphere') resembled an inflated and slightly cross-eyed bullfrog. It had no external air supply. Instead, it carried oxygen tanks and chemicals to absorb excess carbon dioxide. For the trials they couldn't afford a ship with a winch strong enough to lift the five-ton sphere, so Barton melted it down and cast another one that was half the weight and with thinner walls.

Twenty-eight attendants were needed on the surface to tend to the sphere and its communications. The crew worked fine, but the ship, the *Ready*, was anything but. A crewman gazing over the side saw a fish swim up to the hull and vanish inside. With none of her pumps working, she had to run for land before she sank. It would have been unfortunate if the sphere had descended only to be followed by its mother ship.

Before attempting dives to record depths Barton decided to lower the bathysphere unmanned. Imagine his dismay when, on its return to the surface, water was dripping from the closed hatch. They gingerly unbolted the hatch. With a terrifying scream it shot across the deck like a shell from a howitzer and gouged a winch ten metres away. The pressure of the water had squashed all the air inside the sphere into a tiny bubble. When the pressure was released, the bubble instantly expanded to its original size and thrust the water out.

After all the seals had been restuffed, the sphere was ready to be lowered into the ocean – with Beebe and Barton inside.

They entered through the narrow hatch and huddled together on the cold, hard floor of the sphere. The internal diameter of their metal cell was only 137 centimetres. The manhole cover weighed 181 kilograms. It clanged into place, sliding over huge steel bolts. Then enormous nuts were screwed on and banged tight with hammers. Beebe thought of the Edgar Allan Poe story in which the victim is slowly bricked up behind a wall. They were lowered over the side, suspended on a steel hawser two and a half centimetres thick. Both men began to breathe conservatively and converse in whispers.

The initial dives did not go smoothly. At 180 metres down Beebe announced that: 'Only dead men have sunk deeper than this.' As if to prove him correct, water began to leak in round the hatch. Barton suggested that they abort the dive and ask to be hauled up. Beebe thought not. He didn't want to perturb those on deck. Meanwhile, the thick electric cable was being pushed through its seals by the pressure and was coiling menacingly around Barton. By the time they reached the surface they had shipped over nineteen litres of water and Barton was ensnared by more than four metres of serpentine cable.

They tested the emergency-light signals. If anything went wrong and the telephone died, a tiny light would at least indicate they were still alive. On one dive the telephone did fail and their spirits plummeted, for the human voice had been their link with the world above.

At a depth of 485 metres the sphere began to pitch like a balloon in a cyclone. Both of the men clouted their heads against the steel inner surface. For a terrible moment they thought the hawser had snapped and they were tumbling into the abyss. But it was just the heaving of a big sea far above.

At their maximum depth Beebe admired the aquatic illuminations. A pebble of light closed on the window and suddenly exploded into sparks. An unknown luminescent creature had hit the window, setting off a flash of underwater fireworks. He would never forget these living illuminations in the darkness of the icy depths.

Later, on a three-hour dive, literally at the very end of their tether, they reached 920 metres, with the sphere's window holding back over seventeen tonnes of pressure. They had penetrated ten times deeper into the ocean than anyone before them. Beebe could not dismiss the thought of their instant death should the fused-quartz porthole fracture.

Barton's worries were slightly different. He had calculated that the hawser should be able to take the strain, but had doubts that the ship's winch could haul up the combined weight of the sphere and the hawser. The steam boilers powering the winch were working well above their rated pressure and were wheezing like an asthmatic. If the winch and its donkey engine were pulled out of gear with each other, the cable would unwind at a terrifying speed and the sphere would plunge towards the bottom. Barton tried to look on the bright side. At least they would have a very long time to make observations.

Beebe observed several species new to science, but the creatures of the deep were so fantastic that many of his discoveries were discounted at the time. During the Second World War, the sphere was sent on a secret mission to study the effects of depth charges for the US Navy. Barton manufactured an improved sphere, his 'eyeball on a string', and penetrated to a record depth of 1,368 metres (4,500 feet).

Beebe's account of their adventure thrilled the public. One avid reader was Auguste Piccard, an engineer and Professor of Physics at the University of Brussels. He was the epitome of a Hollywood eccentric professor, with hair like Einstein's in an electrical storm. Piccard always wore two watches. He thought that three would be even better as he could then use their average time.

Piccard's mind dwelt on the deficiencies of the bathysphere. Its problems arose from being suspended from a ship. The weight of the hawser limited the sphere's depth penetration. When almost fully unwound on Beebe's deepest dive the weight of the cable was double that of the sphere. Being tethered to the surface also meant that the bathysphere couldn't be directed. If an unknown creature slid past the tiny window, it was glimpsed and then it was gone. Piccard envisaged an independent submersible that could move around to explore the landscape and seek out the fauna.

The metal sphere was obviously a suitable pressure-proof container for the personnel. Piccard's brilliant idea was to sling it beneath a large float containing petrol. The petrol supplied buoyancy, not fuel. It is lighter than water and only slightly compressible. To counteract its buoyancy there would be hoppers full of iron pellets, and two compartments in the float that could be flooded with seawater to make the submersible sink. To slow or halt the descent some iron pellets could be dropped. If lots were liberated the submersible would return to the surface. It would have a round window made from a cone of perspex flat at both ends. Perspex is flexible and doesn't shatter like glass or quartz.

He would call it the bathyscaphe (meaning 'deep boat' and pronounced *bathee skaff*). It could go far deeper than a submarine and cruise around on the ocean bottom using its two small propellers.

Piccard's ambition was to revolutionise oceanography by enabling the scientists to visit anywhere in the oceans. In 1948 the abyssal deeps were *terra incognita*. The largest feature on our planet is a continuous chain of mountains called the Mid-Ocean Ridge. It covers a quarter of the Earth's surface and makes the Himalayas look like an outbreak of zits on a schoolboy's face. Yet until the 1950s we had no idea that it existed.

Our knowledge of deep-sea fauna was also perfunctory. It was said that all the samples taken from the abyssal depths would fit into a single warehouse. They had been collected by blindly dragging a dredge or dropping samplers to take tiny bites out of the sea-floor mud. Imagine that you are drifting over a fogbound London in a balloon and let down a net to trawl the unseen street below. It might 'catch' some cigarette butts, empty beer cans and the regurgitated remains of a doner kebab. How representative would that be of life in the city beneath? Well, perhaps that was a bad example.

Piccard secured funds from the Belgian research council, *Fondes National de la Recherche Scientifique*, so the first bathyscaphe was christened FNRS-2. (Why '2' will become apparent in the next chapter). The sea trial took place in 1948 off the coast of West Africa. It dived unmanned, programmed to release ballast at a predetermined depth and return to the surface. It plumbed 1,398 metres but rose too quickly and the resulting reverberation caused the sphere to leak. The fragile float was also damaged while being towed in a rough sea.

Piccard was soon at work on an improved model. Jacques Cousteau called it the most wonderful invention of the century and the French navy took over the project with Piccard as adviser. The problem was that the navy wouldn't take advice from a non-naval man, an academic dreamer. Auguste Piccard withdrew from the team and built his own submersible. His son Jacques, an economist by training, raised the cash. A new sphere was forged at the Krupp works in Germany, an Italian petroleum company donated the petrol and the Italian navy supplied the support ship. The bathyscaphe *Trieste* was launched in August 1953.

Unlike the original bathyscaphe, the *Trieste* had an access shaft down through the centre of the float to allow the crew to get in from the conning tower when the bathyscaphe was afloat. It reduced the time they had to spend in the sphere. This was important as the crew would be Auguste as observer and Jacques as pilot, both of whom were very tall. The inner diameter of the sphere was only two metres before all the equipment lined its walls. Jacques was almost two metres tall (six feet, five inches). Presumably he had to be folded to get inside.

The sea trials produced a few scares. While deep below, choking smoke filled the sphere. It was just a wire that had shorted, but it was frightening and unpleasant. An unexpected problem was revealed when they released some ballast to slow their descent. The iron pellets followed them down and landed on the *Trieste*'s deck. When they shed more ballast to rise from the bottom, the bathyscaphe didn't budge. Only when they dropped *all* the pellets did it manage to ascend.

The *Trieste* was built to be a tool for oceanographers. To determine whether it was a suitable platform for underwater experimentation several scientists temporarily installed their equipment in the sphere to study sunlight penetration into the sea, sound transmission, and to observe animal behaviour. All were impressed. Seventy hours were spent experimenting at depths down to 300 metres.

The only hairy moment was when the bathyscaphe came to rest on a narrow ledge. It gave way and the *Trieste* slid down a mud slope, triggering an avalanche. Jacques shed ballast to no effect. Anxiously he released another load of pellets and the *Trieste* eased upwards.

The US Office of Naval Research realised the importance of better hydrographic knowledge as submarines were venturing deeper and marauding further afield. The US Navy adopted the Piccards, and the *Trieste* was shipped to San Diego. After struggling for years on a shoestring budget, the project was at last fully funded. The collaboration worked well under the sympathetic command of Lieutenant Don Walsh.

In 1951 the British research vessel *Challenger II* had surveyed the Marianas Trench in the Pacific and found the deepest place in the world's oceans, the Challenger Deep. It was almost 11,000 metres (36,000 feet) down. The US Navy decided it was a challenge they couldn't resist. If the capability existed, they *must* plumb the deepest abyss. But could the *Trieste* do it? Auguste agreed that it *could* be done, but at that immense pressure the margin of safety on the sphere would be small. It could cause a catastrophic implosion. Under pressure hollow objects explode inwards and an implosion can be just violent as an explosion.

Jacques thought it was a diversion from the scientific research, but record breaking is seductive and this was the greatest record of all. He agreed to the dive and insisted on being the pilot. The other crewman would be Don Walsh. A stronger sphere was made, plus a bigger float to lift it.

Test dives began in November 1959. They touched bottom at 5,472 metres, a new world record. As they ascended there were two violent explosions. Had the sphere failed? They surfaced as fast as they could and got out of the sphere. It had been cast in three parts: a central ring and two capping pieces. The epoxy glue that held them together had failed and drops of water were trickling in. Fortunately the water pressure pushed the parts together. As they couldn't be reglued, two metal rings with gaskets underneath were clamped over the seams.

One can hardly blame the crew for feeling nervous. To descend into the abyss imprisoned in a sealed vessel is to become hypersensitive to the slightest creak or ping. Submersibles tend to gurgle and grumble, but the sounds that the *Trieste* uttered were not whispers. When squeezed, it complained.

On one dive a loud implosion sounded like the float splitting. They thought they were dead men, but it was only a camera case that had crumpled. At 7,000 metres there was a series of implosions because holes had not been drilled in hollow metal stanchions to let the water get inside. The air-filled tubes couldn't withstand the water pressure.

Despite the attentions of numerous technicians, there were several equipment failures. The echo-sounder was faulty. It could have been serious had they hit the bottom at speed, thinking they were less deep than they really were. The valve for releasing small amounts of petrol to trim the bathyscaphe was damaged while towing the *Trieste* from the US base on Guam. En route the sea had also carried away the phone, the current meter and the device for measuring rate of descent.

On January 23rd 1960 they located the exact position of the Challenger Deep, but the wind was getting up. The swell was so high that attempting to board the *Trieste* was dangerous. Before they even left the surface the pressure was on. The world's press had arrived and were waiting, two typing fingers poised, to report the great attempt.

Timing was critical. If they delayed any longer the *Trieste* would not return until after sunset and she was a small, inconspicuous vessel to find in the dark. At 8.15 a.m. Piccard and Walsh took a last look at the sky and climbed down into the sphere. As soon as the hatch was sealed, the access shaft was flooded and the *Trieste* began to descend. It was a relief to leave the angry waves behind.

As they got deeper the petrol in the float was compressed and seawater filled the space, so the bathyscaphe became progressively heavier and was falling as fast as a lift in a skyscraper. It grew colder and colder. Their clothes were drenched with condensation and perhaps a little nervous sweating. Outside, the water temperature was 1°C.

The pit of the Challenger Deep was only one and a half-kilometres across and they had no idea how much they might have drifted laterally. They could miss it completely or, worse still, collide with its walls – a frightening prospect.

They were now passing beyond the abyss into what oceanographers call the hadal ('hellish') region. A disconcerting dribble of water was entering the sphere. The pressure on every square centimetre of the perspex port was 1.25 tonnes. Suddenly a heart-stopping implosion shook the sphere. This was a major, perhaps even a fatal failure. They waited, holding their breath. Nothing happened. They exchanged nervous glances, then resumed their descent.

The bottom seemed to rise up to meet them. What they thought was a flatfish slid idly by, oblivious of the occasion. It is now believed that it was a sea cucumber, not a fish. Either way, there was life down there. They stayed twenty minutes in the deepest hole in the world at 10,883 metres, almost seven miles below the surface.

Piccard was concerned that as they rose and the petrol expanded, by the laws of physics it would cool dramatically. The petrol would be all right but the pipe that allowed water in and out of the float to keep its volume constant might freeze. If it did, the float could explode and they would plunge to their deaths. It didn't.

The dive had lasted almost nine hours. On the ascent they discovered that the implosion they had felt was the cracking of a window at the bottom of the access shaft. The shaft was their only way out and it was flooded. If, when they got to the surface, they could not clear the shaft they would be imprisoned in the sphere until the *Trieste* was towed the 322 kilometres back to Guam – not a pleasant prospect in rough seas. They had only condensation to drink and chocolate bars for sustenance. They needn't have fretted. Compressed air displaced the water in the shaft and the window had held firm.

The *Trieste* returned to her scientific duties until she was called to a more sombre task. In 1963 the USS *Thresher* was the newest and most advanced nuclear submarine in the world. On 10 April she was on trials off the coast of New England. The mother ship received a radio message: 'Experiencing minor difficulties . . . Am attempting to blow.' Four minutes later there was a garbled transmission that mentioned 'test depth'. The test depth is the vessel's lowermost diving limit. She was heading for the bottom 2,400 metres below.

At such a depth the steel hull would crumple like paper and water would jet in, destroying everything. There was no chance of anyone surviving. The *Trieste* was rushed from her base in San Diego. She was hastily equipped with the latest cameras and clawed arms for picking up whatever she found. During ten dives, some lasting six hours or more, she discovered a vast field of debris: mangled pipes, buckled steel plates and articles of protective clothing worn in the reactor room. It was the vandalised graveyard of a hundred and fifty men and a mighty submarine humbled by the ocean.

The tragedy stimulated the production of submersibles that could rescue survivors from the depths at which modern submarines operate. Money poured into underwater technology. But, long before, Auguste Piccard had built two bathyscaphes, the most radical submarines ever envisaged, and had done so on a ridiculously small budget. He couldn't afford to equip them with an echo sounder or sufficiently large batteries. Even shedding the iron ballast was a financial worry at $600 at time. To cap it all, he risked his life by going down in them and took his son with him. He merely wished to demonstrate that it was safe.

Piccard did all this because he *had* to. How could he ignore his destiny when the objective was so worthwhile?

Arne Zetterström's gravestone.

High, Fast and Hazardous

'Eyeballs in, eyeballs out' — One consequence of rapid acceleration and abrupt deceleration

Auguste Piccard was sixty-nine when he dived in the bathyscaphe. He had already been immortalised as the lanky professor with feral hair in the adventures of Tintin. Being ambidextrous he could draw a different diagram with each hand simultaneously. Piccard was also bidirectional. He had been to the bottom of the ocean and had also risen into the stratosphere in a balloon. It takes little imagination to realise that the bathyscaphe was just a balloon that sank. As with a balloon the float was an envelope to provide lift, and ballast was dropped to slow or halt the descent.

Auguste had served in the lighter-than-air service during the First World War. Over a decade later he became interested in cosmic rays — the remnants of the Big Bang at the birth of the universe. These rays are nuclear particles that come from outer space. In their passage through the Earth's atmosphere they are modified and lose energy by colliding with other particles. At ground level they contribute little to a person's exposure to radiation but, at the altitude where aircraft fly, cosmic rays become the predominant source of radiation. Auguste wanted to study 'virgin' rays before they were modified. To do this he had to monitor them in the stratosphere.

Venturing to such high altitudes was a risky business, as other researchers had found. In 1862 James Glaisher, a distinguished meteorologist, accompanied by a celebrity balloonist called Henry Coxwell, ascended from Wolverhampton gasworks. The giant balloon had been specially commissioned for scientific research. Glaisher took seventeen scientific instruments to measure humidity and temperature. At an altitude of 8,850 metres his vision became too blurred to read his instruments. Soon his limbs and neck were becoming paralysed and he could no longer speak. They were still rising. Cox climbed up to try to vent hydrogen from the balloon so they could descend, but his hands froze to the metal ring above the basket. He managed to pull the cord with his teeth and saved their lives. They had risen to 11,278 metres – higher than anyone had been before.

On the descent Glaisher recovered and resumed his observations. Undeterred by the adventure he made twenty-eight further ascents. Glaisher was able to show that the higher they went in the atmosphere, the less moisture was present and that the air temperature did not decline uniformly with altitude. The stratosphere was as turbulent as the sea.

In 1875 another ill-fated aerial expedition took off from Paris. It carried three engineers and scientists set on examining the upper atmosphere. Gaston Tissandier, Joseph Sivel and Theodore Croce-Spinelli ascended in a balloon. As they rose they took their pulse and breathing rates. Sivel's heart was beating at almost double the normal rate. At 7,500 metres Sivel suggested they should go higher and the others agreed. They released some ballast and busied themselves taking readings from their barometers, thermometers and spectroscope. The balloon rose rapidly to 8,600 metres and they lost consciousness. Tissandier and later Croce-Spinelli awoke briefly and were so befuddled that they both released more ballast.

When Tissandier came round an hour and a half later, they were at 6,000 metres and falling. Both his colleagues were dead from a lack of oxygen.

In preparation for the flight they had spent time in a low-pressure chamber in the laboratory of Paul Bert who was a world authority on pressure. It convinced them of the importance of breathing oxygen. It would enable them to go even higher than they had planned. They took with them three balloons, about the size of beach balls, filled with seventy per cent oxygen. Bert wrote to them stressing that it was nowhere near enough. They each had only six minutes' supply for a flight on which almost two and a half hours would be spent at altitudes where it was essential. It was too late to get more. The short supply led them to keep it until it was absolutely necessary. By the time Tissandier felt the need for oxygen he was too far gone to reach the mouthpiece.

Gradual oxygen depletion is an insidious killer. As Tissandier wrote, 'One feels an inner joy . . . one becomes indifferent and thinks no more of the perilous situation.' Slow depressurisation in aircraft has led to bizarre accidents in which planes on autopilot have flown on, with all the crew and passengers comatose, until the fuel runs out.

Piccard knew that the answer for his balloon was to have a pressurised cabin with its own oxygen supply. This was long before airplanes were pressurised. For him to study cosmic rays the cabin had to be non-magnetic and electrically neutral. Aluminium was the answer. The only factories that knew how to fashion aluminium were breweries. They assembled large panels to make vats. So brewery engineers shaped three large pieces of aluminium and welded them together to make a sphere. The walls were only 3.5 millimetres thick.

There were two manholes to allow access, with hatches on the inside. To make them airtight the hatch had to be bigger than the manhole. When Piccard came to inspect the finished sphere he noticed that the two hatches were lying on a bench. He pointed out that there was now no way of getting them into the sphere. A workman huffed and heaved and failed to push one through the hole. Yet on Piccard's next visit the hatches were in place. He never discovered how they had done it.

There were also legal problems. For safety reasons the ballast for balloons had to be either sand or water. Piccard had lead shot. He listed it as 'lead sand' and all was well. To ensure that released shot would not injure people on the ground, Piccard stood at the bottom of a fifty-metre chimney under a rain of lead shot.

With the weight of the sphere, and because the air in the stratosphere is so thin, the hydrogen balloon had to be ten times bigger than a conventional hot-air balloon. It was over thirty-four metres in diameter. The project was funded by the Belgian research council and the balloon bore the council's initials: FNRS. That is why much later the first bathyscaphe, funded by the same agency, was designated FNRS-2.

In May 1931 the giant balloon was inflated with hydrogen. A mischievous wind dislodged the sphere from its transporter,which made a tiny hole in the sphere that would endanger the lives of Piccard and his fellow scientist, Paul Kipfer. They had not given the signal to release the moorings when Kipfer saw the top of a factory chimney drifting past.

At an altitude of four kilometres Piccard was concerned to find that the pressure inside and outside the sphere was the same. There was a hiss of escaping gas. A wicker basket would have been as airtight. If they didn't plug the hole, the mission would have to be aborted. Fortunately Piccard had anticipated this development and filled the hole with some 'gunk' that he had prepared earlier. At the second attempt the hissing stopped and the aeronauts enjoyed the reassuring silence.

Within half an hour they were at fifteen kilometres, penetrating into the sombre indigo sky of the stratosphere. After dropping ballast to rise further, they tried to vent a little hydrogen to control their ascent. The rope that allowed them to release hydrogen had jammed. When they tried to twist it free, it snapped. They were now unable to descend and were condemned to drift out of control until their oxygen ran out.

When they didn't land at the scheduled time, the newspapers, with their appetite for bad news, reported:

PICCARD BALLOON DRIFTS HELPLESSLY ABOVE ALPS
Scientist feared dead

Not dead, but becoming increasingly nervous. Piccard bumped into their large barometer and its mercury poured out to pool on the floor of the sphere. Mercury can eat through aluminium. They hoped that the coating of paint wasn't chipped. Then Piccard had a brilliant idea. He connected a length of tubing to a spigot that allowed access to the outside and the vacuum of space sucked up the mercury and spat it out.

At higher altitude the walls of the sphere became so cold that condensation inside the sphere turned to thick frost. It was like being trapped in a cave of ice. When the sun came up, it snowed. As the temperature rose they became parched and neither of them fancied drinking the soup of oil, mercury and water on the floor. Piccard made some water by pouring liquid oxygen into a metal cup and allowing it to evaporate, forming a layer of frost on the outside. It was perfectly drinkable once it had warmed from -212°C.

Piccard had painted one side of the sphere white and the other black so that by rotating it he could have either the heat-absorbing or heat-reflecting side facing the sun. Unfortunately, the rotating mechanism didn't work.

The balloon had risen to 15,781 metres. Then, in the chill of the evening, the hydrogen in the envelope shrank and the balloon began to fall, slowly at first, then faster. They didn't dare to slow it down by dropping ballast for fear that the balloon would rise out of control. The dying sun caught the curve of the balloon and observers on the ground discovered a new crescent moon.

Below them were the sharp pointed arêtes of the Alps. The sphere bounced like a pinball over a glacier with crevasses deep enough to swallow it whole. Spotting a snowfield beyond, Piccard pulled the strap of the balloon's ripping panel to release all the hydrogen. The huge envelope collapsed around them. It made an ample counterpane under which they could sleep beneath the stars. Next morning Piccard woke in a panic. He thought the whisper of a distant waterfall was the hiss of escaping air.

Because Piccard and Kipfer spent so much time coping with emergencies they made only a perfunctory measurement of cosmic radiation. On his return Piccard immediately began work on a new sphere to return to the stratosphere. In August 1932 he ascended to a record altitude, only sixty metres short of seventeen kilometres. His sphere is now on display at the Science Museum at Wroughton in Wiltshire.

He showed that cosmic radiation was more intense in the stratosphere but not so great as had been suspected. Even at the maximum altitude he reached, relatively few high-energy particles penetrated into the sphere. He confirmed that astronauts could survive voyages into the high stratosphere and perhaps beyond, into outer space.

Auguste Piccard's twin brother Felix later designed high-altitude balloons for the US Air Force that reached 30,000 metres. In 1999 his grandson Bertrand and Brian Jones were the first balloonists to circle the world. But Auguste wasn't interested in records, not even his own.

Many records, such as the sound barrier, have been broken in the course of research for military purposes. During the Second World War German jet fighters just zipped away when a flight of Mustangs appeared. The US Air Force was determined never to be outpaced again.

At the end of the war a young fighter ace was posted to what is now called Wright Field, a US Air Force base in Ohio. Chuck Yeager became an Assistant Maintenance Officer. He and another ex-fighter pilot, Bob Hoover, were allowed to take up any of the assortment of aircraft on the base. Their forays were not without incident. Hoover had twenty or so aerial scrapes. One time, after his engine had failed, he deliberately bounced off the top of a truck in order to clear a fence. Not surprisingly, the resident test pilots considered them to be undisciplined cowboys. In contrast, they were college types, cool and careful.

Wright Field was the centre for testing a new generation of super-fast aircraft. Yeager was upgraded to test pilot just as the Air Force took receipt of a unique research aircraft, the Bell X-S1. The X stood for experimental and the S for supersonic. It cost $6 million to develop and it was top secret. The orange fuselage was shaped like a bullet and was propelled by four rockets called 'Black Betsy'. It was built for one purpose: to smash through the sound barrier. It would not be easy and would definitely be dangerous. Strange things happened to planes that dared to approach the speed of sound, Mach 1* (1,216 kilometres per hour at sea level).

* Because the speed at which sound travels varies with altitude, air speed is often measured in relation to the speed of sound (Mach 1) at the height the plane is flying. This scheme was devised by Ernst Mach, an Austrian physicist.

In 1943 Miles Aircraft in England was at the forefront of aeronautical innovation. One of the Miles brothers was the father of Mary Miles, who married the ill-fated Peter Small. The brothers produced the M.52, an experimental plane with 'special edition' Whittle engines that were far more powerful than other jet engines of the time. It was expected to reach 2,600 kilometres per hour. Its sharp-edged wings, designed to slice through the turbulence when close to Mach 1, earned it the nickname the 'Gillette Falcon'. Engineers from the Bell Aircraft Company came to England to consult with the Miles brothers. Perhaps as a consequence, the Bell X-S1 developed ultra-sharp leading edges on its wings like those of the M.52. Unfortunately, the M.52 never got a shot at the sound barrier; the British government cancelled the contract.

De Havilland, another British Company, built a revolutionary swept-wing, tail-less research plane powered by four gas turbines. Wind-tunnel tests of the D.H.108 revealed some instability, as with other tailless aircraft, but in more than a hundred test flights she performed well and reached 0.89 Mach in level flight. The owner's son Geoffrey de Havilland was piloting the D.H.108 when, in a fast dive, it became violently unstable and disintegrated. Geoffrey died of a broken neck before he hit the ground. His brother had already been killed in a mid-air collision.

Chuck Yeager knew from experience that as an aircraft approached Mach 1 it began to shake and the controls could 'freeze'. Perhaps the sound barrier was more than just a name. Some aeronautical engineers believed that as the pressure built up in front of a speeding aircraft the shock wave at Mach 1 might be like hitting a wall, which neither pilot nor plane would survive.

The top brass at Wright Field asked for volunteers to pilot the X-1 (the 'S' had been dropped to hide the fact that it was attempting to go supersonic). Both Yeager and Bob Hoover stepped forward. They were not short on courage. During their training as fighter pilots their squadron lost thirteen pilots in six months. They had survived, and daily dogfights with enemy planes made them fatalistic. But they never used the word 'crash'. Instead, anyone who ploughed into the ground was said to have 'bought the farm'. The RAF too said 'he bought it' or, more enticingly, 'he went for a Burton'. Yeager figured that they were accustomed to sorties on which they never knew what might happen next. Experimental flights were no different.

To the astonishment of all the higher-ranking test pilots, Yeager, a mere captain, was chosen to be the principal test pilot for the X-1 flights, with Bob Hoover as his understudy. The other pilots scoffed that the powers above had selected the two most expendable pilots. No one believed they had a hope in hell.

In fact, Yeager was chosen because his commanding officer believed that if it could be done Yeager was the man to do it. He had a instinctive 'feel' for every aircraft he flew. It was as if he became one with the machine.

The worst chore was training. Yeager hated the giant centrifuge in which they frequently – and literally – went for a spin. It exposed them to forces many times greater than the Earth's gravitational pull and took them way beyond nausea. Even the bravest pilots broke out in a cold sweat when strapped in the centrifuge for another stint of life without blood in their head. There were also endless sessions in a cold low-pressure chamber in conditions similar to those to be found at 21,000 metres. They tested pressurised suits and on one occasion Hoover had difficulty breathing and his face turned purple. They had forgotten to attach his oxygen supply.

A corset maker constructed the flight suits. When flying back from a fitting at the factory Chuck and Bob's plane was struck by lightning and almost went down. To be killed on a routine flight while waiting to pilot the most frightening brute ever invented would have been the ultimate irony.

Chuck chose Jack Ripley as his flight engineer for the experimental flights. Ripley had to be kept away from the explosive fuels as he was a chain-smoker and his shirts were full of burn holes from fallen cigarette ash. He was well weathered. Although only in his late twenties, he looked much older.

Chuck knew that Ripley was not just hot on the theory of aeronautics, he was an immensely practical man. When a pilot made an emergency landing on a small airstrip Yeager and Ripley went to retrieve the plane. It was stranded because, loaded with fuel, it would need a much longer runway to take off. Ripley calculated the minimum amount of fuel it would need to get back to base. He then paced out the runway and hammered in a stake to mark the spot where Yeager should fire the jet booster on take-off. Ripley assured him that he had three metres to spare. Yeager plunged down the runway and lifted off – with three metres to spare.

At last Yeager got to meet the X-1. Even in its hangar it was chained down like a wild beast that had to be restrained. On his first flight the X-1 would carry no fuel. He would practise gliding down and landing. Even when powered, the X-1 would have to glide home because all of its several tonnes of fuel would be burnt within 4.2 seconds.

The X-1 didn't take off. It was carried aloft slung beneath a modified B-29 bomber. At the designated altitude Yeager had to climb down a ladder in the bitingly cold wind and slide into the cockpit. He put on his helmet. As they hadn't supplied him with one, he had cut down the leather helmet usually worn by a tank commander.

Yeager heard the pop of the release and dropped rapidly out of the bomber's shadow into dazzling sunlight. He was on his own. The X-1 handled like a dream and he enjoyed the silent slide towards the ground.

In August 1947 he made his first X-1 powered flight. The rockets ran on alcohol and liquid oxygen. They were meant to be explosive. When the X-1 carried fuel for a test flight the entire base was shut down. If there was an explosion

Yeager would have a ringside seat. The cockpit was filled with non-flammable nitrogen so he was reliant on an independent oxygen supply. With tanks of liquid oxygen at -182° C right behind him Yeager felt chilled. Apprehension also contributed. If the aircraft had faults they would reveal themselves here, when he and the X-1 were alone together, 12,000 metres above the ground.

There was no bail-out option. The door was at the side of the cockpit and if he managed to jump he would be instantly bisected by the razor-edged wing. Nonetheless, they gave him a parachute. It made a good cushion. The only outcomes were that he would become a hero or a dead hero. He couldn't lose. So he fired the first rocket.

Flames leapt six metres behind the plane and Yeager felt as if a truck had just rammed him from behind. He was *travelling*. At 300 metres below the mother ship he ignited a second rocket and was thrust up to 0.7 Mach – and this was on half-power. He made a barrel roll in celebration and the engines cut out. He would get a rollicking when he landed. He had strict instructions not to fire the remaining two rockets so he ditched the remaining fuel before gliding home.

On his sixth test flight he hit 0.86 Mach and the aircraft began to shudder as if it were running on a cobbled street. The controls became sluggish. The seventh flight was worse. At 0.94 Mach the hand controls failed to respond. If, as predicted, the nose pitched on breaking the sound barrier, and he couldn't correct it, it would be curtains. The project was about to be abandoned when Ripley had an idea. He fitted a motor that would alter the configuration of the tailplane independent of the hand controls. Many years later an airliner lost control and the pilot used the wing engines to lift the nose and the third engine on the tail to move the nose down. His ingenuity saved the lives of over two hundred passengers. Ripley was trying a different way to do something similar. But would it work at the speed of sound?

It worked on the next flight at 0.988 Mach, but the windscreen frosted over. Yeager was blind and with no navigation instruments to get him home. Hoover in the chase plane talked him down and he made his most gentle landing. Wiping the windscreen with hair shampoo solved the frosting problem.

While preparing for his attempt to break the sound barrier, Yeager fell off a horse and broke a couple of ribs. He was in pain, but with tight bandaging and painkillers he thought he would be able to fly. The main problem was that he couldn't lean over to lock the cockpit door. Ripley sawed off a piece of broom handle that allowed him to secure the door.

On 14 October 1947 Yeager's ribs had a bumpy rocket ride. When he fired the final rocket he noticed that the needle on the speed gauge had gone off the scale. At that moment he vanished from sight and there was a loud explosion. Observers thought that Yeager was lost, but it was the first-ever sonic boom. He was travelling at Mach 1.07.

There was no announcement from the Department of Defense. They wanted to keep it secret until they developed a supersonic fighter. Yeager continued as a test pilot and his very next flight in the X-1 was the scariest. When he clicked the switch to fire the rockets, nothing happened. All he could do was to ditch the fuel and glide down, but without power he couldn't open the valves to expel fuel. He remembered there was a manual control for venting and used it. Without a working gauge he had no idea how fast it was bleeding out or how much fuel was left in the tanks. The flimsy landing gear of the X-1 was designed to take the weight of the aircraft. It would collapse under the extra burden of any significant amount of fuel left in the tanks and an explosion would be inevitable. Yeager delayed landing for as long as possible and made the most nervous touchdown of his life. His luck held.

What was Yeager's reward for his skill and daring? Civilian test pilots got substantial danger money, even when working for the military. Yeager got a bar on his Distinguished Flying Cross and later an array of awards including a peacetime Congressional Medal of Honor. But he was banned from making any money from publicising his exploits and was not promoted from captain to major until seven years later.

It had been clear for a while that the increasing speed of fighter aircraft made it difficult for the pilot to escape from a disabled airplane. Clambering out of the cockpit was impossible when the plane was travelling at 800 kilometres an hour or more. He had to be forcibly ejected. Experiments with dummies indicated that the pilot might survive. Miles Aircraft patented an ejector seat as early as 1939, and the Martin Baker company tested a modern ejector seat in 1945.

The principle was that an explosive charge would simultaneously remove the cockpit's canopy and blast the pilot and his seat clear of the plane. In the Martin Baker factory there was a test rig that thrust dummies aloft up a steep chute. Bernard Lynch, a fitter from the shop floor, volunteered to be the first person to ride the rig.

The tests showed that a force of 25 G (twenty-five times the gravitational pull of the Earth) applied 'gradually' over a tenth of a second would hoist the pilot up at a velocity of eighteen metres per second to clear an aircraft flying at 800 kilometres per hour. And it did this without concertinaing his spine. The ejector seat was installed in a Meteor III jet fighter and test pilot Bryan Greensted survived ejection.

To protect the pilot, the explosive charge was activated by pulling a heavy canvas sheet over his head. When it reached his waist the seat blasted off. The force of his ejection ensured that his hands pulled the hood tightly over his head, preventing any violent flexing of the neck.

The pilot's problems were not over when he departed the aircraft. Four years before Yeager arrived at Wright Field another self-experimenter was putting himself at risk. Lieutenant Colonel William Lovelace was director of the Aero-Medical Laboratory. He considered the parachutist's plight.

Within a minute of bailing out at altitude, a pilot went unconscious and couldn't pull the ripcord to release his parachute. Lovelace developed an oxygen breather to keep him alive until he dropped low enough to breathe the air.

To prove it worked in practice he bailed out of a B-17 bomber at an altitude of 12,200 metres. It was the first time he had jumped out of a plane. He plummeted in free fall and then released his parachute. He had assumed that the jolt of it opening would be about 4 G, but at this altitude the jolt delivered an almost instant deceleration of 33 G. His body suddenly became thirty-two times heavier. It knocked him unconscious. It also tore away his glove at an altitude where eyelids freeze shut. Having survived such an enormous G-force he ricked his back on landing where forces involved were a paltry 3 to 4 G. Like Yeager, he was awarded the DFC for his pains.

To avoid an excessive jolt pilots free-fall until they reach terminal velocity (thirty-three metres per second) before opening their chute. Thanks to Lovelace's experience parachutes that opened automatically were developed.

Rapid deceleration is what does the damage to the occupants when an aircraft or vehicle crashes. Also, when a pilot ejects, he suffers abrupt linear deceleration the instant be emerges from the speeding plane. Colonel John Stapp, a flight surgeon, decided to explore the limits of human tolerance of deceleration. He was the ideal person. In addition to his medical degree he had a PhD in biophysics. He gathered data on types of restraining harness and attended post-mortems to see how the harnesses performed in accidents. Perhaps he could improve their design.

To decelerate Stapp first had to accelerate. What better than a rocket-propelled sled? A heavy-duty railway track was laid near the Holloman Air Force Base in New Mexico. It was over 1,000 metres long. The sled would not run *on* the tracks; they would merely keep it on a straight line. The sled *Sonic Wind* was manufactured by the Northrop Aircraft Company. It had twelve rockets that delivered 24,490 kilo-grams of thrust in just 0.07 seconds. There were no conventional brakes: a bucket scoop underneath the sled dug into a trough of water between the tracks. It was the equivalent of hitting a brick wall.

Thirty preliminary runs with a dummy in the sled gathered useful data. Stapp had calculated the forces involved. Now he had to find out what they felt like. He had been ordered to supervise the project and decided that *he* should be the main guinea pig. He was relieved that he was asked to save lives rather than take them. He later allowed others to volunteer but made sure they were aware they were going to get hurt. His volunteers included pilots, flight surgeons, a medical technician and two harness makers.

Stapp adopted a rigorous regime prior to a test run. A measure of his courage was that before every run he contemplated the fearful consequences if anything were to go wrong. Then he fired the rockets.

Stapp rode the sled twenty-two times. Each run was faster than the previous one. He tried to keep his hands well inside the fairing. Twice before, a straying hand had caught the wind and his wrist had fractured instantly. He concentrated on keeping alert so that he would remember the details of the experience for his report. His aim was to determine the 'point of beginning injury'. He could only do so by going *beyond* that point.

His twenty-second run on 12 December 1954 would be his fastest. Before the rockets were ignited Stapp's heart rate soared and his spirits sank. His protective gear was just a harness, a helmet and a rubber bite to prevent him from guillotining his tongue with his teeth. The rockets fired and within five seconds he was travelling at over 1,000 km per hour (632 mph). He could overtake a .45 calibre bullet. There was no windscreen so he was buffeted by the wind. Sand bit through his flight suit to pincushion his skin. His eyes were being compressed into his skull. He withstood a force of 40 G for twenty seconds – a very long time to endure such stress. As the blood supply to his eyes was disrupted his vision dimmed and he went through a blackout (everything black with eyes wide open), followed by a red-out. The scoop hit the water and the sled stopped in one and a half seconds. Stapp's eyes ballooned out of their sockets. Had his retina detached he would have been blinded.

When Stapp began his experiments experts believed that

a human could not withstand more than 9 G. To go beyond that was courageous. To go so far beyond was extraordinary. In 1958 one of Stapp's colleagues survived a sled run that subjected him to 82.6 G. His name was Eli Beeding, a name too close to 'He lies bleeding' for comfort. He survived because he was wearing Stapp's improved harness and sitting up so the forces hit him in the back and chest. Had he been lying down it would have been a different matter. The US Air Force came up with the idea that it might be safer if ejector seats propelled the pilot forward rather than upwards, just as a high diver enters the water head first. Stapp decided this was too dangerous for a man to try so a heavily anaesthetised monkey rode the sled lying down. It did not survive. In that position the blood was rammed up into the skull, bursting blood vessels and destroying tissue.

The worst that Stapp suffered was extensive bruising and concussion, and an abdominal hernia, as well as fractured ribs, wrists and coccyx. He also developed chronic vertigo if he shut his eyes and tried to balance. He refused any compensation for his injuries because they were merely routine hazards for which he received standard flight pay. He treated the experiments as active duty. They were just as hazardous as being on the battlefield and servicemen expected to be in the line of fire.

The speed of progress in aviation was astonishingly fast. The first flight of the Wright brothers' biplane on December 17th 1903 lasted twelve seconds and a jogger on the beach at Kitty Hawk would have overtaken it. It reached an altitude of two metres. Only sixty years later *Concorde* made its maiden flight and was soon taking paying passengers across the Atlantic faster than the speed of sound. These advances would not have been possible but for the experiments of many brave men.

Stapp became chief of the Air Medical Laboratory, and was later placed on permanent loan to the Department of Transport for research into car restraints. In the same year that he had survived an abrupt stop from 1,000 kilometres per hour almost 40,000 American motorists had died from crashes at forty kilometres an hour or slower.

Stapp became an important advocate for increased car safety and devoted himself to making car crashes more survivable. This required more human deceleration experiments for him and others. In polite conversation the term 'biomechanics' was used to refer to the violent business of determing how much impact people could stand. Stapp boasted that between 1947 and 1970 he and his volunteers had endured over five thousand voluntary human experiments without a single disabling injury or loss of life.

The Society of Automotive Engineers established a foundation in his name for training engineers in safety matters, and the Stapp Car Crash Conferences are held regularly all round the world to report the latest developments in car safety.

The US car industry resisted all change. Their priority was

styling, not safety. The president of General Motors boasted that his company was run by salesmen, not engineers. When seat belts were first introduced into production cars there were those that claimed they would damage drivers more than the collision. But they hadn't ridden the *Sonic Wind*. Thanks to John Stapp and his experiments many hundreds of thousands of people are alive after surviving potentially fatal car accidents.

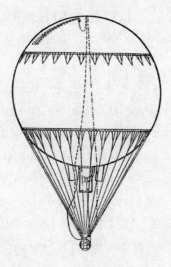

Piccard's giant balloon. The tiny sphere at the bottom housed the crew.

Risky Business

'We know not, and cannot know, where safety ends and danger begins'
— Dr Walter Channing

Life is a series of uncontrolled experiments. We call some of these misadventures adolescence, courting and parenthood. From time to time people take unnecessary risks: darting across the road in front of an oncoming truck, swimming with sharks, jumping from a bridge in the expectation that an elastic rope won't snap, or from a plane in the belief that parachutes always open. Like the pole-vaulter they think only of the leap, never the fall. Pioneers in any field take a leap into the unknown and sometimes it's a risky venture.

In medicine, animal tests became a prerequisite to reduce the risks to patients from new drugs. This followed a scandal in the United States in 1937. Sulfanilamide was the first 'wonder drug' for bacterial infections. In those days there was no requirement for testing new compounds. One brand was sweetened with diethylene glycol, what we now know as anti-freeze. Patients began dying in agony. The (unqualified) chemist of the company swigged a slug of their concoction to demonstrate its safety. Within a day he was dangerously ill. Almost all the 909 litres of elixir that had been distributed were traced and confiscated, but too late to save the hundred and seven recipients who died.

Congress rapidly passed a bill that forced drug companies to prove that a product was safe, and animal testing became the norm. Although we share virtually all our physiology and biochemistry with other mammals, our responses are not always identical. Aspirin is fatal to cats and penicillin kills guinea pigs. Fortunately the biochemists who purified penicillin and carried out the first tests of its effects used mice. Had they chosen guinea pigs, the first antibiotic might never have made it into production.

The regulations governing animal experiments have always been more strict than those concerning experiments on humans, hence Jack Haldane's quip: 'To do the sort of things to a dog that are done to the average medical student requires a licence signed in triplicate by two archbishops.' In an essay entitled 'On being one's own rabbit', Haldane pointed out that rabbits make 'little serious attempt to cooperate with one'. Worse still, 'dumb' animals can't tell you how they feel, and this can be essential in physiological studies. Nor do animals share all our diseases. They are not susceptible to cholera or yellow fever so all the experimentation for those diseases *had* to be done on humans.

Experimenting on others comes naturally to us. Cleopatra was reluctant to commit suicide without first establishing which poisons were most rapid and least unpleasant. So, before plumping for an asp-assisted exit, she is said to have tested a range of products on her handmaidens. She dismissed strychnine as it was not only agonising but also left the deceased with a sardonic smile, which was most unflattering.

In the eighteenth and nineteenth centuries the slums of big cities swarmed with destitute citizens who were seen as appropriate subjects for medical experimentation. With the introduction of smallpox inoculations to protect the recipient from exposure to more virulent strains, inmates of Newgate Prison tested the safety of the procedure, with a pardon for the survivors. The success of this trial led to inoculation becoming a compulsory method of prevention and the term 'conscientious objector' was first used in 1898 to describe those who risked prosecution for refusing to have their children inoculated. When Dr Benjamin Waterhouse introduced vaccination into the USA, he persuaded the Boston Board of Health to conduct a public demonstration. Nineteen volunteers were given cowpox and then, two weeks later, an injection of smallpox. Two 'control subjects' received only smallpox. The trials were 'completely successful', which presumably means that only the controls died.

Small sample sizes (often smaller by the end of the experiment) were a feature of such tests.

In those days few medics had qualms about such experiments because diseases were incubated in the slums from where 'infectious miasmas' arose to plague respectable citizens. It seemed appropriate that the poor should be the vehicle by which medicine advanced.

Sick patients were also readily to hand and in constant supply. The terminally ill were 'obvious candidates' for drug trials. They had nothing to lose, but also little to gain. If they were in the advanced stages of a fatal disease, even the most wondrous of wonder drugs would be unlikely to help them. It was neither a fair trial of the drug's capabilities, nor a proper way to care for the dying.

Doctors did not restrict their treatment to curing the disease from which the patient was suffering. Often they ignored the central tenet of medicine – do no harm. The desire to research overruled the duty of care. A doctor confided that he didn't tell patients they were taking part in experimental trials 'out of consideration for the patient'. One doctor who was spraying deadly germs into the noses of patients admitted: 'They thought I was treating them for nasal congestion.' It was a betrayal of trust: 'Our patients obeyed us gladly. Our zeal led them to respect and trust us. It never occurred to them to enquire whether this zeal was in the interests of treatment or in the interests of science.'

The zeal of surgeons later became apparent in the race to

carry out the first successful heart transplant. Of all the patients receiving hearts up to June 1969, fifty died in less than a month, ninety lasted less than two and a half months and only two survived longer.

Dr Pappworth's scouring of medical journals revealed that even in the 1950s and 1960s patients in Britain and the USA were regularly subjected to risky procedures that contributed nothing to their cure. Such abuses still occur. From 1998 to 2000 over a hundred children in a Catholic care home in New York were subjects in trials requiring high doses of dangerous drugs. Between 1998 and 2003 the General Medical Council in Britain took action in a dozen cases of fraudulent research by general practitioners. They were giving untested drugs to patients without explaining the dangers involved or even informing them that they were taking part in an experimental trial. At least one of the doctors received £100,000 from a pharmaceutical company for conducting the experiment.

Such conduct flouts the rules that govern medical practice. The Nuremberg Code was drawn up following the atrocities wrought by Nazi doctors on prisoners in concentration camps. At the heart of the Code and of others that followed is the necessity to obtain the patients' *informed* consent before subjecting them to any experimental treatment. Even so, in 1954 the secretary of a hospital management board countered a patient's complaint of being operated on without his permission by claiming: 'If a patient comes into hospital . . . he is deemed to agree to receive treatment.' Although nowadays every university and hospital has an ethics committee to assess all research proposals, much depends on how frank the researcher is with the volunteer about the potential discomfort and risk, and how much of the explanation of the procedures can be understood by the average patient.

Barry Marshall, who injected himself with *Helicobacter,* gave as the reason for experimenting on himself: 'I was the only person informed enough to give consent.' Enoch Hale, the first person to be injected with a medicine, also chose self-experimentation because only 'professional men can estimate the inconvenience or risk to which they may be subjected'.

All researchers should ask themselves: 'Would I submit myself to this experiment?' If not, then the experiment should not be attempted. Dr Chauncey Leake, a distinguished pharmacologist who endured several painful self-experiments, was adamant that pharmacologists developing new drugs 'have a moral obligation to try such drugs on themselves . . . before using them experimentally on any other human being'. Enoch Hale stated that for hazardous experiments volunteers should not be used 'even if they would be willing to undergo them'. Most self-experimenters believed that in all conscience they couldn't ask even the most plucky volunteer to swallow live parasites or undergo a painful or dangerous procedure.

Where there is danger, someone must go first. Self-experimenters have greater protection than the non-scientist because of their detailed knowledge of what they are doing. Physiologists and medics are better able to read their symptoms and recognise warning signs during their experiments. Self-experimenters also have a strong incentive to alleviate the stress of any procedure. As John Stapp said of his rocket-propelled sled experiments: 'You can design harnesses and restraints that are far better after you ride with one of your mistakes.'

Scientific researchers are obsessives. Dr William Bean measured the growth of his fingernails daily for thirty years to correlate it with his health. Another American doctor cracked the finger joints on his left hand for fifty years to see if it accelerated arthritis. When this degree of determination is directed at more serious matters it may drive the experimenter to take unwarranted risks. Thomas Brittingham, who injected himself with blood from patients with cancer and leukaemia, became addicted to self-experimentation. He confessed that at the time he had not considered the impact on his family of his possible death.

Auguste Piccard cautioned against impetuosity. He stressed that every potential danger had to be anticipated and every risk assessed. Only then should the scientist proceed.

In reality it is often the unimagined risks that bite. Every self-experiment carries some danger. Otherwise they would be unnecessary. Jack Haldane put these dangers into perspective: 'Others make experiments which are apparently dangerous, but really perfectly safe provided the theory on which they are based is sound. I have occasionally made experiments of this kind and if I had died in the course of one, I should, while dying, have regarded myself not as a martyr, but as a fool.' Roald Amundsen, the polar explorer, put it more succinctly: 'Adventure is just bad planning.'

Many self-experimenters think that it's what the other guy does that's dangerous. Jack Haldane wrote: 'Experiments in which one stakes one's life in the correctness of one's biochemistry are far safer than those of an airplane designer who is prepared to fall a thousand feet if his aerodynamics are incorrect. They are also more likely to benefit humanity.'

Perhaps no action can be entirely altruistic, but the research of these experimenters comes close. Whatever their ambition and ego, they were as likely to be criticised as praised for their efforts. Werner Forssmann, the first to catheterise the heart, was sacked for his pioneering spirit. A few achieved fame, but how many of the pioneers in this book are known to the public? The Curies became famous for their research, not for their courage in continuing to work with radioactive chemicals after they knew they were dangerous. Self-experimenters are not masochistic or suicidal: they are the bold shock troops of research.

It is ethically preferable for John Scott Haldane and his colleagues to have tested the effects of poison gases on themselves, compared to some of the alternatives. In 1943 Australian soldiers volunteered for experiments with mustard gas. They were warned in advance that there was 'a small risk of burns and blisters'. In the event they suffered severe burns all over the body, violent vomiting and headaches, loss of nails and teeth, and developed chronic lung problems. Some died. In the 1950s and early 1960s British servicemen who believed they were involved in research into the common cold were actually exposed to mustard gas and sarin, a deadly nerve gas. Several almost died and one did. The verdict of an inquest held over fifty years later was 'unlawful killing'. Little use was made of the findings of either of these military experiments, whereas Haldane's self-experiments led to the development of the gas mask. If a researcher risks his own life, he makes damn sure that the research has a clear purpose.

Although many pioneers led lives at the outer limits of probability, they coolly endured the dangers and discomforts that their experiments engendered because they believed their mission was worthwhile. The proof that it was often research of the highest standard is evidenced by the number of Nobel Laureates who were self-experimenters.

They took risks to make our lives safer. Jack Haldane said: 'It is occasionally necessary to make experiments which one knows are dangerous, for example in determining how a disease is transmitted. A number of people have died in this way. It is to my mind the ideal way of dying.' Haldane had no desire to die prematurely. His thoughts were noble, not suicidal.

Even some of the sternest critics of experimentation on humans applauded those who experimented on themselves. Sir William Osler, the greatest medical educator of his time, wrote: 'The history of our profession is starred with the heroism of its members who have sacrificed health and sometimes life itself in endeavours to benefit their fellow creatures.'

In a selfish world, society needs such people. We should celebrate them.

Attempting any of the self-experiments described in this book is not recommended.

Recommended Reading

Lawrence Altman's *Who Goes First?*, University of California Press, 1998 is *the* source on medical self-experimentation. It is detailed and based not only on the published literature but also on innumerable interviews with the researchers themselves. Were it not for his meticulous research many of the stories of medical self-experimentation related here would have been lost forever.

Roy Porter's *The Greatest Benefit to Mankind*, Fontana Press, 1999 is an excellent history of medicine that provides the background to many of the experiments.

The numerous articles from *New Scientist* that are listed in the bibliography cannot be bettered for interesting and readable accounts of the non-medical pioneers.

For experiments on people other than oneself, see S. Lederer's *Subjected to Science*, Johns Hopkins University Press, 1995.

The mentality of the self-experimenter has never been described with more clarity and wit than by J. B. S. Haldane in 'On being one's own rabbit' in his delightful collection of essays *Possible Worlds*, reprinted by Transaction Publisher U.S. in 2001.

Credits

The author would like to thank all those responsible for giving permission to reproduce pictures and extracts from copyright material.

Altman, L. K., *Who Goes First?* reproduced by kind permission of the University of California Press

Austin, W., engraving entitled *A night watchman disturbs a body-snatcher*, Wellcome Library, London

'*Physicians of the Utmost Fame*' by Hilaire Belloc (© Hilaire Belloc) is reproduced by permission of PFD (www.pfd.co.uk) on behalf of The Estate of Hilaire Belloc.

Reprinted by permission from Macmillan Publishers Ltd: NATURE, Bishop of Birmingham, November 29th, 1930, © 1930

Engraving of the *Gravitator*, reprinted from The Lancet, vol.12 issue 302, Blundell, Observations on the transfusion of blood, p.321, © 1829, with permission from Elsevier

Buckland, F., engraving of him with a porpoise, from *Curiosities of Natural History*, *Third series*, Richard Bentley & Son, 1873.

Churchill, W.S., *The Second World War*, Vol. 2 *Their Finest Hour*, Cassell, 1949. Reproduced with permission of Curtis

Brown Ltd, London on behalf of The Estate of Winston Churchill. Copyright © Winston S. Churchill

Editorial, *ECT in Britain*, The Lancet, 1981, 28th November, 1207–1208, reprinted with permission from Elsevier

Editorial, *The Mosquito Hypothesis*, Washington Post, November 2nd 1900, by permission of PARS International Corporation, New York

Thanks to Craig Ferreira, for permission to use his pithy phrase on the possibility of being 'munched' by sharks

George, A, *Hard to swallow*, New Scientist, Dec 9th 2006, with permission of New Scientist Syndication (RBI-UK)

Extracts from books of Professor Hans Hass with permission, © Hans Hass-Archive (HIST)

Leake, C., *Technical triumphs and moral muddles*, Annals of Internal Medicine, 67, suppl. 7, 1967, with permission, Medical Reprints

Mellanby, K., *Human Guinea Pigs*, Merlin Press, London, 1973 reproduced by permission of The Merlin Press

Diagram of balloon from A. Piccard, *Au Fonde des Mers en Bathyscaphe*, 1954, copyright B. Arthaud, Éditions Flammarion

Throckmorton, P., *The Lost Ships*, Jonathan Cape, 1976, by kind permission of Paula Throckmorton Zakaria and Lucy Throckmorton

Illustration of Arne Zetterstrom's grave based on photograph by Dr John Bevan

Haldane, J.B.S., *Possible Worlds*, reproduced by permission of Transaction Publishers © 2001.

While every effort has been made to secure permissions, I apologise for any apparent negligence on my part and undertake to make any necessary corrections in future editions.

Bibliography

He came, he sawed, he chancred

Bondeson, J., 'Three remarkable specimens in the Hunterian Museum', in *A Cabinet of Medical Curiosities*, I. B. Tauris & Co. Ltd., London & New York, 1997, 186–215

British Broadcasting Corporation, *'US undertakers admit corpse scam'*, http://news.bbc.co.uk/2/hi/americas/6064692.stm 19 October 2006

Dickenson, D., *Body Shopping. The Economy Fuelled by Flesh and Blood*, One+World, Oxford, 2008

Editorial, 'The Edinburgh Murders', in *The Lancet*, 3 January 1829, 433–438

Fox, D., 'Can masturbating each day keep the doctor away?' in *New Scientist*, 19 July 2003, p.15

Harris, P., 'They replaced stolen bones with pipes; organs with rags – 1,077 corpses carved up by illegal body snatchers who sold their "harvest" on the donor market', in *The Observer Magazine*, 2 April 2004, 20–27

Hollingham, R., *Blood and Guts. A History of Surgery*, BBC Books, London, 2008

Hunter, J., 'A Treatise on the Venereal Disease', 1786, extract in L. Clendening, *Source Book of Medical History*, Dover Publications Inc., New York, 1942, 488–499

Hunter, W., *The Anatomy of the Human Gravid Uterus Exhibited in Figures*, Baskerville, Baker & Leigh, Birmingham, 1774, reprint Sydenham Society 1851

Iserson, K.V., *Death to Dust: What Happens to Dead Bodies*, 2nd edition, Galen Press, Tucson, 2001

Jones, J.H., *Bad Blood. The Tuskegee Syphilis Experiment*, new and expanded edition, The Free Press, New York & London, 1993

Love, B., *Encyclopaedia of Unusual Sex*, Greenwich Editions, London, 1999

Moore, W., *The Knife Man*, Bantam Press, London, 2005

Moreno, D., *Undue Risk. Secret Laboratory Experiments on Humans*, Routledge, New York & London, 2001

Norton, T., 'Living Proof', in *Times Educational Supplement. Curriculum special*, 2001, 6–7

Palmer, J.F., ed., *The Works of John Hunter*, Longman, Rees, Orme, Brown, Breen, London, 4 volumes, 1835–1837

Phillips, S., 'The return of the body-snatchers', in *Times Higher Educational Supplement*, 26 March 2004, p. 22

Quist, D., *John Hunter (1728–1795)*, William Heinemann Medical Books, London, 1981

Revill, J. & D. Campbell, 'Calls grow for organ transplant revolution' and 'One transplant kidney can save my son's life', in *The Observer*, 13 January 2008, pp.1, 3, 28–30

Richardson, R., *Death, Dissection and the Destitute*, 2nd edition, Phoenix Press, London, 2001

Richardson, R. & B. Hurwitz, 'Donors' attitudes towards body donation for dissection', in *The Lancet*, 29 July 1995, 277–279

Sanders, C., 'Why low body count is fatal for anatomy', in *Times Higher Educational Supplement*, 6 June 2003

Sawday, J., *The Body Emblazoned*. Routledge, London, 1996

Simmons, J.G., 'John Hunter/ Beginning of scientific medicine and surgery', in *Doctors and Discoveries*, Houghton Mifflin Co., Boston & New York, 2002, 140–144

Southey, R., 'The surgeon's warning' in *Poems 1799*, reprint, Kessinger Publishing, 2004

Stubbs, G., *The Anatomy of the Horse*, 1776, new edition with a modern paraphrase by J. C. McClunn assisted by C.W. Ottaway, Heywood Hill, London, 1938

Wilkinson, C., editor, *The Observer Book of the Body*, Observer Books, London, 2008

Sniff it and see

Altman, L. K., *Who Goes First?*, University of California Press, Berkeley, 1998

Ayer, W., 'Account of an eye-witness to the first public

demonstration of ether anaesthesia at the Massachusetts General Hospital, October 16, 1846', in L. Clendening, *Source Book of Medical History*, Dover Publications Inc., New York, 1942, 372–373

Boon, M., *The Road to Excess*, Harvard University Press, Cambridge, USA, 2002

Booth, F., H.J. Bigelow & R. Liston, 'Surgical operations performed during insensibility, produced by the inhalation of ether', in *The Lancet*, 2 January 1847, 5–8

British Broadcasting Corporation, '*Medical Mavericks I*', BBC 4 TV, 4 February 2007

Collins, P., 'Poe's cure for death', in *New Scientist*, 13 January 2007, 50–51

Crowther, J. G. 'Humphry Davy 1778–1829', in *British Scientists of the Nineteenth Century*, vol. 1, Pelican Books, Harmondsworth, 1940, 15–81

Davy, H., '*Researches, Chemical and Philosophical; Chiefly Concerning Nitrous Oxide, or Dephlogisticated Nitrous air, and its Respiration*', J. Johnson, London, 1800

Editorial, 'Administration of chloroform to the Queen', in *The Lancet*, 14 May 1853, p. 453

Franklin, J. & J. Sutherland, *If I Die in the Service of Science*, Morrow, New York, 1984

Freud, S. 'Über Coca', in *Centralblatt für die gesamte Therapie*, 2, 1884, 289–314

Hogan, P., 'Soft drink, hard sell', in *The Observer Magazine*, 9 July 2006, 26–29

Holmes, R., 'Davy on the gas', in *The Age of Wonder*, HarperPress, London, 2008, 235–304

Jay, M., *Artificial Paradises*, Penguin Books, London, 1999

Koller, C., 'Historical note on the beginning of local anaesthesia', in *Journal of the American Medical Association*, 90, 1928, 1742–1743

Long, C.W., 1853, 'First Surgical operation under ether', in L. Clendening, *Source Book of Medical History*, Dover Publications Inc., New York, 1942, 356–358

Morton, W.T.G., 1847, 'Remarks on the proper mode of administering ether by inhalation', in L. Clendening, *Source Book of Medical History*, Dover Publications Inc., New York, 1942, 366–372

Nowak, R., 'Nitrous oxide is no laughing matter', in *New Scientist*, 11 August 2007, p.15

Pain, S., 'This won't hurt a bit', in *New Scientist*, 16 February 2002, 48–49

Pain, S., 'Blissful oblivion', in *New Scientist*, 17 March 2009, 44–45

Prescott, F. 'Discussion on further experiences with curare', in *Proceedings of the Royal Society, Medicine*, 40, 1947, 593–602

Prescott, F., G. Organe & S. Rowbottom, 'Tubocurarine chloride as an adjunct to an anaesthesia', in *The Lancet*, 2, 1946, 80–84

Simpson, J.Y., 'On a new anaesthetic agent, more efficient than sulphuric ether', in *The Lancet*, 20 November 1847, 549–550

Southey, C.C., *Southey, Life and Correspondence*, Longman, London, 1849–1850

Stratmann, L., *Chloroform. The Quest for Oblivion*, Sutton Publishing, Stroud, 2005

Wilkinson, C., editor, *The Observer Book of Money*, Observer Books, London, 2007

Trials and tribulations

Agin, D., *Junk Science*, Thomas Dunne Books, St Martin's Press, New York, 2006

Altman, L. K., *Who Goes First?* University of California Press, Berkeley, 1998

Anon., 'Vioxx settlement', in *New Scientist*, 17 November 2007, 6–7

Battacharya, S & A. Coghlan, 'One drug, six men, disaster . . .', in *New Scientist*, 25 March 2006, 10–11

Blacow, N.W., editor, *Martindale. The Extra Pharmacopoeia*, 26th edition, The Pharmaceutical Press, London, 1972

Brookes, M., *Extreme Measures. The Dark Visions and Bright Ideas of Francis Galton*, Bloomsbury Publications, London, 2004

Channel 4 TV, *Dispatches: The Drug Trial That Went Wrong*, 28 September 2006

Clarke, M., 'Clinical trials and tribulations', in *Times Higher Educational Supplement*, 24 March 2006, p. 23

Defoe, D., *A Journal of the Plague Year*, 1772, reprint Dent, London, 1966

Fielding, H., *The History of Tom Jones*, 1749, Folio Society edition, London, 1959

Gilmour, J., *British Botanists*, Collins, London, 1946

Jewson, N., 'Medical knowledge and the patronage system in eighteenth-century England', in *Sociology* 8, 1974, 369–385

Jewson, N., 'The disappearance of the sick man from medical cosmology 1770-1870', in *Sociology* 10, 1976, 225–244

Lamont-Brown, R., *Royal Poxes and Potions. The Lives of Court Physicians, Surgeons and Apothecaries*, Sutton Publishing Ltd., Stroud, 2001

MacKie, R. & J. Revill, 'Trial and error', in *The Observer*, 19 March 2006, 23–25

Marshall, M., 'So many questions and so little justice', in *The Observer*, 24 December 2006, 16–17

Moore, W., *The Knife Man*, Bantam Press, London, 2005

Motluk, A., 'Occupation: lab rat', in *New Scientist*, 25 July 2009, 41–43

Murrell, W., 'Nitro-glycerine as a remedy for angina pectoris', in *The Lancet*, 1879, 80–81, 113–115, 151–152, 225–227

Norton, T., 'Living Proof', in *Times Educational Supplement. Curriculum special*, 2001, 6–7

Pain, S., 'Mrs Carlill lays down the law', in *New Scientist*, 14 January 2006, 50–51

Porter, R., *Quacks. Fakers and Charlatans in English Medicine*, Tempus Publishing Ltd., Stroud, 2000

Porter, R., *Blood and Guts*, Penguin Books, London, 2003

Revill, J., 'Drug trial firm knew of risk', in *The Observer*, 9 April 2006

Revill, J., R. McKie & A Hill, 'Drug chief defends tests on volunteers', in *The Observer*, 19 March 2006, p. 2

Shetty, P., 'Plight of the human guinea pig', in *New Scientist*, 11 July 2009, p. 48

Thompson, S., *New Guide to Health or Botanic Family Physician*, New Edition, Simpkin, Marshall & Co., London, 1849

Lovely Grubs

Bompas, G.C., *Life of Frank Buckland*, Smith, Elder & Co., London, 1885

Brock, A.J., 'The Reverend William Buckland, the first palaeoecologist', in *Biologist*, 40 (4), 1993, 149–152

Brookes, M., *Extreme Measures. The Dark Visions and Bright Ideas of Francis Galton*, Bloomsbury Publications, London, 2004

Buckland, F., *Curiosities of Natural History*, First-Fourth Series, Richard Bentley & Son, London 1873–1874

Buckland, F., *Notes and Jottings from Animal Life*, Smith, Elder & Co., London, 1886

Burgess, G.H.O., *The Curious World of Frank Buckland*, John Baker, London, 1967

Evans, H.M., *Sting Fish and Seafarer*, Faber & Faber, London, 1943

Furlow, B., 'The freelance poisoner', in *New Scientist*, 20 January 2001, 30–33

Galton, F., *The Art of Travel*, 1872, Phoenix Press edition, London, 2000

Gardner-Thorpe, C., 'Who was Frank Buckland?', in *Biologist*, 48 (4), 2001, 187–188

Halsted, B.W., *Dangerous Marine Animals*, Cornell Maritime Press, Cambridge, Maryland, 1959

Hopkins, J., *Strange Foods*, Periplus Editions (HK) Ltd., Hong Kong, 1999

Lockwood, S.J., 'Buckland professors and dining clubs', in *Biologist*, 48 (5), 2001, p. 200

Newman, C., '12 Toxic tales', in *National Geographic*, May 2005, 4–31

Pain, S., 'Bat out of Hell', in *New Scientist*, 10 January 2004, 50–51

Ritvo, H., *The Platypus and the Mermaid and other Figments of the Classifying Imagination*, Harvard University Press, Cambridge Massachusetts, 1997

Scherschel, J.J., 'Puffer', in *National Geographic*, August 1984, 4–31

Spinney, L., 'The killer beans of Calabar', in *New Scientist*, 28 June 2003, 48–49

Wilkinson, C., editor, *The Observer Book of Food*, Observer Books, London, 2008

A diet of worms

Altman, L. K., *Who Goes First?* University of California Press, Berkeley, 1998

Anon, 'Diet of worms protects against bowel disease', in *New Scientist*, 10 April 2004, p. 8

Anon, 'Malaria vaccine', in *New Scientist*, 13 December 2008, p. 7

Barlow, C.H., 'Experimental ingestion of the ova of *Fasciolopsis buski*; also the ingestion of adult *F. buski* for the purpose of artificial infestation', in *Journal of Parasitology*, 8, 1921, 40–44

Barlow, C.H. & H.E. Meleney, 'A voluntary infection with *Schistosoma haematobium*', in *American Journal of Tropical Medicine*, 29, 1949, 79–87

Barnes, R.S.K., P. Calow & P.J.W. Olive, *The Invertebrates: a New Synthesis*, Blackwell Scientific Publications, Oxford, 1988

Blacow, N.W., (ed.), *Martindale. The Extra Pharmacopoeia*, 26th edition, The Pharmaceutical Press, London, 1972

Bondeson, J., 'The Bosom Serpent', in *A Cabinet of Medical Curiosities*, I. B. Tauris Publishers, London & New York, 1997, 26–50

Buchsbaum, R., *Animals Without Backbones*, Vol. 1, Pelican Books, Harmondsworth, 1951

Carlin, J., 'It's the world's deadliest disease, killing more than 900,000 a year in Africa alone. But can Bill Gates' dollars create a vaccine that could save a continent's children?', in *The Observer Magazine*, 17 February 2008, 26–34

Connor, S.J., 'Malaria in Africa: the view from space', in *Biologist*, 46 (1), 1999, 22–25

Geddes, L., 'A diet of worms could keep MS at bay', in *New Scientist*, 20 January 2007, p. 8

Harris, E. & L. Middleton, 'The discrete charm of nematode worms', in *New Scientist*, 22/29 December 2007, 70–71

Harris, J.E. & H.D. Crofton, 'Famous animals: *Ascaris*', in *New Biology*, 27, 1958, 109–127

Johnson, M.L., 'Malaria, mosquitoes and man', in *New Biology*, 1, 1945, 96–109

Johnson, M.L., 'Famous animals: The tapeworm', in *New Biology*, 7, 1949, 113–123

Lindsay, S. & R. Hutchinson, 'Will malaria return to the UK?' in *Natural Environment Research Council News*, Spring 2002, 22–23

Kaplan, M., 'Benefits of parasites', in *New Scientist*, 11 July 2009, p. 43

McKie, R, 'Now the doctors say parasitic worms are good for you', in *The Observer*, 13 May 2001, p. 6

Mercer, J.G. & L.H. Chappell, 'Appetite and parasite', in *Biologist*, 47 (1), 2000, 35–40

Pearce, F., 'Set free to kill again', in *New Scientist*, 6 October 2007, 58–59

Snow, K., 'Could malaria return to Britain?' in *Biologist*, 47 (4), 2000, 176–180

Zuk, M., 'The joy of parasites', an interview in *New Scientist*, 23 June 2007, 44–45

The desire for disease

Altman, L. K., *Who Goes First?* University of California Press, Berkeley, 1998

Anon, 'Gut reaction', in *New Scientist,* 8 October 2005, p. 7

Anon., 'Yellow fever alert', in *New Scientist,* 8 March 2008, p. 7

Boese, A., 'The whacko files – 9: The vomit-drinking doctor', in *New Scientist,* 3 November 2007, 54–55

Chadwick, E., *The Report of an Enquiry into the Sanitary Condition of the Labouring Population of Great Britain,* 1842, reprinted by Edinburgh University Press, 1965

Dickens, C., *Bleak House,* 1852–1853 reprinted by Penguin, London, 2003

Editorial, 'The Cholera', in *British Medical Journal,* 1856, 848–849

Editorial, 'The mosquito hypothesis', in *Washington Post,* 2 November 1900 (Quoted in Altman, 1998).

George, A., 'Hard to swallow', an interview with Barry Marshall, in *New Scientist,* 9 December 2006, p. 53

Gorgas, W.C., 'Results in Havana during the year 1901 of disinfection for Yellow Fever', in *The Lancet,* 6 September, 1902, 166–169

Howard-Jones, N., 'Gelsenkirchen typhoid epidemic of 1901, Robert Koch and the dead hand of Max von Pettenkofer', *British Medical Journal,* 1973, 103–105

Kelly, H.A., *Walter Reed and Yellow Fever,* new and revised edition, McCluer Phillips & Co., 1907

Latta, T., 'Malignant cholera', in *The Lancet,* 2 June 1832, 274–277

Laurence, B.R., 'The discovery of insect-borne disease', in *Biologist,* 36 (2), 1989, 65–71

Lax, A., *Toxin. The Cunning of Bacterial Poisons*, Oxford University Press, 2005

Lederer, S., *Subjected to Science. Human Experimentation in America before the Second World War*, Johns Hopkins University Press, Baltimore, 1995

Longmate, N.R., *King Cholera. The Biography of a Disease*, Hamish Hamilton, London, 1966

McCaw, W.D., *Walter Reed. A Memoir*, Water Reed Memorial Association, 1904

Marshall, B.J., 'Attempt to fulfil Koch's postulates for pyloric *Campylobacter*', in *Medical Journal of Australia*, 142, 1985, 436–439

Morris, R.D., *The Blue Death. Disease, Disaster and the Water We Drink*, OneWorld, Oxford, 2007

Pettenkofer, M., 'On cholera with reference to the recent epidemic in Hamburg', in *The Lancet*, 1892, 1182–1185

Reed, W. 'The propagation of Yellow Fever: Observation based on recent researches', 1901, in L. Clendening, *Source Book of Medical History*, Dover Publications Inc., New York, 1942, 479–484

Reed, W., J. Carroll & A. Agramonte, 'The etiology of yellow fever. An additional note', in *Journal of the American Medical Association*, 36, 1901, 431–440

Reed, W., J. Carroll, A. Aristides & J.W. Lazear, 'The etiology of yellow fever. A preliminary note', in *Philadelphia Medical Journal*, 148, 1900, 790–796

Ross, R., 'The role of the mosquito in the evolution of the malarial parasite', in *The Lancet*, 20 August 1898, 488–489

Simmons, J.G., 'Louis Pasteur. The germ theory of disease', 18–23; 'Robert Koch. Foundations of bacteriology', 24–28; 'John Snow. Field epidemiology begins at the Broad Street Pump', 162–4, in *Doctors and Discoveries*, Houghton Mifflin Co., Boston & New York, 2002

Snow, J., *On the Mode of Communication of Cholera*, 1824, in L. Clendening, *Source Book of Medical History*, Dover Publications Inc., New York, 1942, 468–473

Snow, J., 'The mode of propagation of cholera', in *The Lancet*, 16 February 1856, p.184

Snow, J., 'Cholera and water supply in the south districts of London', in *British Medical Journal*, 17 October 1857, 864–865

Warren, J.R. & B. Marshall, 'Unidentified curved bacilli on gastric epithelium in active chronic gastritis', in *The Lancet*, 4 June 1983, 1273–1275

Weyers, W., *The Abuse of Man*, Ardor Scribendi Ltd, New York, 2003

The disease detectives

Abraham, C., 'West knows best', in *New Scientist*, 21 July 2007, 35–37

Albrink, W.S., S.M. Brooks, R.E. Biron & M. Kopel, 'Human inhalation anthrax. A report of three fatal cases', in *American Journal of Pathology*, 36, 1960, 457–471

Aldous, P., 'Shaky mental history was no bar to anthrax work', in *New Scientist*, 23 August 2008, p. 12

Anon, 'Another martyr to yellow fever', in *Journal of the American Medical Association*, 91, 1928, 107–108

Anon, 'Bacteriologist dies of meningitis', in *Journal of the American Medical Association*, 106, 1936, p. 129

Anon, 'Anthrax mix-up', in *New Scientist*, 19 June 2004, p. 5

Anon, 'Ebola accident', in *New Scientist*, 11 April 2009, p. 4

Asthana, A., 'Inside Ebola's zone of death', in *The Observer*, 16 December 2007, p. 29

Bright, M., & S. Cooper, 'Walter Mitty life of anthrax terror suspect', in *The Observer*, 1 June 2003, p. 20

British Broadcasting Corporation, 'Huge US payout over anthrax case', in http://news.bbc.co.uk./2/hi/americas/7478722.stm 27 June 2008

British Broadcasting Corporation, 'Pressure killed anthrax suspect', in http://news.bbc.co.uk./2/hi/americas/7538373.stm 1 August 2008

British Broadcasting Corporation, 'Scientist "lone anthrax attacker"', in http://news.bbc.co.uk./2/hi/americas/7545398.stm 6 August 2008

Cunningham, W., 'The work of two Scottish medical graduates in the control of woolsorters disease', in *Medical History*, 1976, 169–173

Geddes, L., 'Animal lab mishaps go unreported', in *New Scientist*, 22/29 December 2007, p.11

Guillemin, J., *Anthrax. The Investigation of a Deadly Outbreak*, University of California Press, Berkeley, 2001

Guillemin, J., *Biological Weapons*, Columbia University Press, 2005

Hammond, E., 'Keep biodefence honest', in *New Scientist*, 6 October 2007, p. 24

Johnson, K.M., P.A. Webb, J.V. Lang & F. A. Murphy, 'Isolation and partial characterisation of a new virus causing acute haemorrhagic fever in Zaire', in *The Lancet*, 12 March 1977, 569–571

Lax, A., *Toxin. The Cunning of Bacterial Poisons*, Oxford University Press, 2005

Lederer, S., *Subjected to Science. Human Experimentation in America before the Second World War*, Johns Hopkins University Press, Baltimore, 1995

McKenna, M., *Beating Back the Devil*, Free Press, New York, 2004

MacKenzie, D., 'Lab slip-up could trigger next flu epidemic', in *New Scientist*, 23 April 2005, p. 11

MacKenzie, D., 'Marburg virus found in fruit bats', in *New Scientist*, 1 September 2007, p. 14

MacKenzie, D., 'The hunter and the doomsday virus', an interview with Bob Swanepoel, in *New Scientist*, 3 November 2007, 56–57

MacKenzie, D., 'Behind the 2001 anthrax attacks', in *New Scientist*, 28 February 2009, p. 13

Meselson, M., J. Guillemin, M. Hugh-Jones, A. Langmuir, I. Popova, A. Shelokov & O. Yampolskaya, 'The Sverdlovsk anthrax outbreak of 1979', in *Science*, 266, 1994, 1202–1208

Peters, C.J. & M. Olshaker, *Virus Hunter*, Anchor Books New York, 1997

Spinney, L., 'Welcome to Fort Plague', in *New Scientist*, 19 April 2008, 44–45

Townsend, M., 'Terrorists try to infiltrate UK's top labs', in *The Observer*, 2 November 2008

Virgil, *The Georgics of Virgil*, Cape, London, 1943

Walker, D.H., O. Yampolska [sic] & L.M. Grinberg, 'Death in Sverdlovsk: What have we learned?' in *American Journal of Pathology*, 144, 1994, 1135–1140

That unhealthy glow

Alexander, F.W., 'A victim to science. X-ray martyr', in *The Lancet*, 22 January 1910, p. 267

Anon, 'Power freak electrocuted', in *Irish Independent*, 13 September 2006, p. 25

Anon, 'Safer and cheaper MRI scanners', in *New Scientist*, 17 November 2007, p. 27

Bourke, J., *Fear. A Cultural History*, Virago, London, 2005

Brown, G.I., *Invisible Rays. A History of Radioactivity*, Sutton Publishing, Stroud, 2002

Duchenne, G.B., *The Mechanism of Human Facial Expression*, 1862, Cambridge University Press, 1990

Cameron, D.E., 'Psychic driving', in *American Journal of Psychiatry*, 112, 1956, 502–509

Cameron, D.E., 'Production of differential amnesia as a factor in the treatment of schizophrenia', in *Comprehensive Psychiatry*, 1, 1960, 26–34

Collins, P., 'Nothing but a ray of light', in *New Scientist*, 8 September 2007, 68–69

Crooks, 'A life history with X-rays', in *The Journal of the Radiological History and Heritage Charitable Trust*, 2000, 11–38

Dalyell, T., 'Eric Voice. Chemist who volunteered as a guinea pig', in *The Independent*, 19 October 2004

Editorial, 'ECT in Britain: A shameful state of affairs', in *The Lancet*, 28 November 1981, 1207–1208

Freund, L., 'Ein mit Röntgen-Strahlen behandelter Fall von Naevus pigmentosus piliferous', in *Wiener Medizinische Wochenschrift*, 47, 1897, 428–434

Goodchild, S., 'Hundreds of patients given shock treatment without their consent', in *The Independent on Sunday*, 13 October 2002, p. 8

Gourlay, K., 'Scientist inhales deadly plutonium for test', in *The Independent on Sunday*, 10 December 2000

Harvie, D.I., *Deadly Sunshine. The History and Fatal Legacy of Radium*, Tempus Publishing Ltd., Stroud, 2005

Jones, R. & O. Lodge, 'The discovery of a bullet lost in the wrist by means of Roentgen ray', in *The Lancet*, 22 February 1896, 476–477

Lemov, R., *World as Laboratory. Experiments with Mice, Mazes and Men*, Hill & Wang, New York, 2005

Mackay, C., 'The magnetisers', in *Extraordinary Popular Delusions and the Madness of Crowds*, 2nd edition, 1852, Wordsworth Editions 1995, 304–345

Maple, E., *Magic, Medicine and Quackery*, Robert Hale, London, 1968

Martland, H.S., 'Occupational poisoning in manufacture of luminous watch dials', in *Journal of the American Medical Association*, 9 February 1929, 466–473

Meyer, H.W., *A History of Electricity and Magnetism*, Burndy Library, Norwalk, Connecticut, 1972

Moore, W., *The Knife Man*, Bantam Press, London, 2005

Mould, R.F., 'Early medical X-rays'; 'Marriage and X-rays'; 'The medico-legal significance of X-rays in the first year after their discovery'; 'Interview with Pierre Curie and ninety years later', in *Mould's Medical Anecdotes*, omnibus edition, Institute of Physics Publishing, Bristol & Philadelphia, 1996, pp. 39–44, 49, 262–269 & 421–425

Newton, D., 'Eric Voice' in *The Independent*, 29 October 2004

Polednak, A.P., A.F. Stehney & R.E. Roland, 'Mortality among women first employed before 1930 in the U.S. radium dial-painting industry', in *American Journal of Epidemiology*, 107, 1978, 179–194

Porter, R., *The Greatest Benefit to Mankind. A Medical History of Humanity from Antiquity to the Present*, Fontana Press, London, 1999

Porter, R., *Quacks. Fakers and Charlatans in English Medicine*, Tempus Publishing Ltd., Stroud, 2000

Röntgen, W.C., 'Uber Eine Neue Art von Strahlen', 1895, in

L. Clendening, *Source Book of Medical History*, Dover Publications Inc., New York, 1942, 666–675

Rowntree, C., 'Development of X-ray carcinoma', in *The Lancet*, 20 March 1909, 821–824

Simmons, J.G., 'W. C. Röntgen. The discovery of X-rays', in *Doctors and Discoveries*, Houghton Mifflin Co., Boston & New York, 2002, 102–104

Smith, A., *The Mind*, Penguin Books, Harmondsworth, 1985

Stott, J.R.R., 'Vibration', in Rainford, D.J. & D.P. Gradwell, (eds), *Ernsting's Aviation Medicine*, 4th edition, Hodder Arnold, London, 2006, 231–246

Watson, L., *Supernature*, Hodder & Stoughton, London, 1974

Found to be wanting

Altman, L. K., *Who Goes First?* University of California Press, Berkeley, 1998

Anson, G., 1853, *A Voyage Round the World in the Years 1740, 1741, 1742, 1743, 1744*, reprint of the 1st edition compiled by R. Walker, S. Jones & B. Robins, Oxford University Press, 1974

Bown, S.R., *Scurvy. How a Surgeon, a Mariner and a Gentleman Solved the Greatest Mystery of the Age of Sail*, Thomas Dunne Books, St Martin's Press, New York, 2003

Crandon, J.H., C.C. Lund & D.B. Dill, 'Human experimental scurvy', in *New England Journal of Medicine*, 223, 1940, 353–369

Dickman, S.R., 'The search for the specific factor in scurvy', in *Perspectives in Biology and Medicine*, 24, 1981, 382–395

Freyer, J., 'How we all became vitamin junkies', in *Daily Express*, 1 October, 2004, 44–45

Herbert, V., Experimental nutritional folate deficiency in man', in *Transactions of the Association of American Physicians*, 75, 1962, 307–320

Hopkins, G., 'Diseases due to deficiencies in diet', in *The Lancet*, 8 November 1913, 1309–1310

Hough, R., *Captain James Cook*, Coronet Books, Hodder & Stoughton, London, 1994

Hughes, R.E., 'James Lind and the cure for scurvy. An experimental approach', *Medical History*, 19, 1975, 342–351

Lederer, S., *Subjected to Science. Human Experimentation in America before the Second World War*, Johns Hopkins University Press, Baltimore, 1995

Lloyd, C., ed., *The Health of Seamen. Selections from the Works of Dr James Lind, Sir Gilbert Blane and Dr Thomas Trotter*, Publications of the Navy Records Society, volume 107, 1965

Lund, C.C. & J.H. Crandon, 'Human experimental scurvy and the relation of vitamin C deficiency to post-operative pneumonia and to wound healing', in *Journal of the American Medical Association*, 116, 1941, 663–668

Siler, J.F., P.E. Garrison & W.J. MacNeal, *Pellagra. First Progress Report of the Thompson–McFadden Pellagra Commission*, 1914, 1–109

Various, 'Pellagra in England: An account of four recent cases', in *British Medical Journal*, 1913, volume 1, 1–12

Something in the Blood

Altman, L. K., *Who Goes First?* University of California Press, Berkeley, 1998

Blundell, Dr, 'Observations on the transfusion of blood. With a description of his Gravitator', in *The Lancet*, 13 June 1829, 321–324

Brittingham, T.E. & H. Chaplin, 'The antigenicity of normal and leukemic human leukocytes', in *Blood*, 17, 1961, 139–165

Davies, H.W., J.B.S. Haldane & E.L. Kennaway, 'Experiments on the regulation of the blood's alkalinity I', in *Journal of Physiology*, 54, 1920, 32–45

Franklin, J. & J. Sutherland, *If I Die in the Service of Science*, Morrow, New York, 1984

Goldstein, J., G. Siviglia, R. Hurst, L. Lenny & L. Reich, 'Group B erythrocytes enzymatically converted to Group O, survive normally in A, B and O individuals', in *Science*, 215, 1982, 168–170

Grant, S.B. & A. Goldman, 'A study of forced respiration: experimental production of tetany', in *American Journal of Physiology*, 52, 1920, 209–232

Haldane, J.B.S., 'Experiments on the regulation of the blood's alkalinity II', in *Journal of Physiology*, 55, 1921, 265–275

Haldane, J.B.S., 'On being one's own rabbit', in *Possible Worlds*, Chatto & Windus, London, 3rd edition, 1945, 107–119

Harrington, W. J., V. Minnich, J.W. Hollingsworth & C.V. Moore, 'Demonstration of a thrombocytopenic factor in the blood of patients with thrombocytopenic purpura', in *Journal of Laboratory & Clinical Medicine*, 38, 1953, 1–10

Harrington, W. J., C.C. Sprague, V. Minnich, C.V. Moore, R.C. Aulvin & R. Dubach, 'Immunologic mechanisms in idiopathic and neonatal thrombocytopenic purpura', in *Annals of Internal Medicine*, 38, 1953, 433–469

Hollingham, R., *Blood and Guts. A History of Surgery*, BBC Books, London, 2008

Le Sage, A.R. 'The Adventures of Gil Blas of Santillane', 1715–1735, extract in L. Clendening, *Source Book of Medical History*, Dover Publications Inc., New York, 1942, 287–296

Nowak, R., 'Blood doesn't always save lives', in *New Scientist*, 26 April 2008, 8–9

Pepys, S., *The Diary of Samuel Pepys 1660–1669*, G. Bell & Sons Ltd, London, 1922

Porter, R., *The Greatest Benefit to Mankind. A Medical History of Humanity from Antiquity to the Present*, Fontana Press, London, 1999

Smith, A., *The Body*, Penguin Books, Harmondsworth, revised edition 1985

Sprague, C.C., W. J. Harrington, R.D. Lange & J.B. Shapleigh, 'Platelet transfusions and the pathogenesis of idiopathic thrombocytopenic purpura', in *Journal of the American Medical Association*, 150, 1952, 1193–1198

Thiersch, J.B., 'Attempted transmission of human leucemia [sic] in man', in *Journal of Laboratory and Clinical Medicine*, 30, 1945, 866–874

Webster, C., 'The origin of blood transfusion. A reassessment', in *Medical History*, 1971, 387–392

A change of heart

Altman, L. K., *Who Goes First?* University of California Press, Berkeley, 1998

Bono, E. de, ed., *Eureka! How and When the Greatest Inventions were Made*, Thames & Hudson, London, 1974

Boyadjian, N., *The Heart and its History, its Symbolism, its Iconography and its Diseases*, Esco Books, Antwerp, 1980

Forssmann, W., 'Sondierung des rechten Herzens', in *Klinische Wochenschrift* 8, 1929, 2085–2087

Forssmann, W., 'Über Kontrastdarstellung der Hohlen des lebenden rechten Herzens und der Lungenschlagader', in *Muenchener Medizinische Wochenschrift*, 78, 1931, 489–492

Forssmann, W., *Experiments on Myself*, St Martin's Press, New York & London, 1974

Grüntzig, A., 'Transluminal dilatation of coronary artery stenosis', in *The Lancet*, 4 February 1978, p. 263

Hollingham, R., *Blood and Guts. A History of Surgery*, BBC Books, London, 2008

Laënnec, R.T.H., *Traité de l'Auscultation Médiate*, 2 vols., J.A. Brosson & J.S. Chaude, Paris, 2nd edition, 1826

Nissen, R., 'Historical development of pulmonary surgery,' in *American Journal of Surgery*, 89, 1955, 9–15.

Pappworth, M.H., *Human Guinea Pigs. Experimentation on Man*, Pelican Books, Harmondsworth, 1967

Porter, R., *The Greatest Benefit to Mankind. A Medical History of Humanity from Antiquity to the Present*, Fontana Press, London, 1999

Simmons, J.G., 'René Laënnec. The physician's new gaze', in *Doctors and Discoveries*, Houghton Mifflin Co., Boston & New York, 2002, 62–66

Smith, A., *The Body*, Penguin Books, Harmondsworth, revised edition 1985

Behind the lines

Adie, K., *Into Danger*, Hodder & Stoughton, London, 2008

Altman, L.K., *Who Goes First?* University of California Press, Berkeley, 1998

Bebb, A.H., 'Direct and reflected explosion waves in deep and shallow water', in *Royal Naval Personnel Research Committee Report*, March 1955, 1–7

Bebb, A.H., H.N.V. Temperley & J.S.P. Rawlins, 'Underwater blast: Experiments and researches by British investigators', in *Admiralty Marine Technology Establishment Report* A M T E (E) R81 401, 1981, 1–69

Bebb, A.H. & H.C. Wright, 'The effect of an underwater explosion on a subject floating on the surface in a submarine escape immersion suit', in *Royal Naval Personnel Research Committee Report*, July 1952, 1–3

Bebb, A.H. & H.C. Wright, 'Underwater explosion blast data from the R. N. Physiological Laboratory 1950– 55', in *Royal Naval Personnel Research Committee Report*, April 1955, 1–7.

Birchall, P., *The Longest Walk. The World of Bomb Disposal*, Arms & Armour Press, London, 1997

Brickhill, P. *The Dam Busters,* Evans Brothers Ltd., London, 1951

Calder, R., *The People's War*, Jonathan Cape, London, 1969

Churchill, W.S., *The Second World War*, Vol. 2 *Their Finest Hour,* Cassell, London, 1949

Elliott, D. H., 'A short history of submarine escape: The development of an extreme air dive', in *South Pacific Underwater Medical Sciences*, 29(2), 1999, 81–87

Fox, B., 'Carry On, Spooks', in *New Scientist*, 24/31 December 2005, 70–71

Hald, J. & E. Jacobsen, 'A drug sensitizing the organism to ethyl alcohol', in *The Lancet*, 1948, 1001–1004

Hunter, C., *Eight Lives Down*, Bantam Press, London, 2007

Kemp, D.J., S.F. Walton, P. Harumal & B.J. Currie, 'The scourge of scabies', in *Biologist*, 49 (1), 2002, 19–24

Mellanby, K., 'The development of symptoms, parasitic infection and immunity in human scabies', in *Parasitology*, 35, 1944, 197–206

Mellanby, K, *Human Guinea Pigs*, Merlin Press, London, 1973

Miller, F.T., *History of World War II*, John C. Winston Co. Ltd., Toronto, 1945

Norton, T. 'Boom! Horace Cameron Wright', in *Stars Beneath the Sea*, Arrow Books, London, 2000, 144–158

Norton, T., 'Living Proof', in *Times Educational Supplement.* Curriculum special, 2001, 6–7

Suffer

Baker, N., 'Decade of decompression, 1897–1908', summary by R. Vallintine in *Historical Diving Times*, 26, 200, 10–11

Behnke, A.R., 'Physiologic investigations in diving and inhalation of gases', in K. R. Dronamraju, ed., *Haldane and Modern Biology*, Johns Hopkins Press, Baltimore, 1968, 267–275.

Boycott, A.E., G.C.C. Damant & J.S. Haldane, 'The prevention of compressed-air sickness' in *Journal of Hygiene*, 8, 1908, 342–441.

Case, E.M. & J.B.S. Haldane, 'Human physiology under high pressure I. Effects of nitrogen, carbon dioxide and cold', in *Journal of Hygiene*, 41, 1941, 225–232

Clarke, R., *J. B. S. The Life and Work of J. B. S. Haldane*. Hodder & Stoughton, London, 1968

Douglas, C.G., 'John Scott Haldane', *Obituary notices, The Royal Society of London*, 1936

Goodman, M., *Suffer and Survive. The Extreme Life of Dr J. S. Haldane*, Simon & Schuster, London & New York, 2007

Haldane, J.B.S., 'On being one's own rabbit', in *Possible Worlds*, Chatto & Windus, London, 1927, 107–119

Haldane, J.B.S., 'Mathematics of air raid protection', in *Nature*, *London* 142, 1938, 791–792

Haldane, J.B.S., *A. R. P.*, Victor Gollancz, London, 1938

Haldane, J.B.S., *Keeping Cool*, Chatto & Windus, London, 1940

Haldane, J.B.S., 'Human life and death at increased pressure', in *Nature*, London, 148, 1941, 458–462

Haldane, J.B.S., 'Life at high pressure', in *Penguin Science News*, 4, 1947, 9–29

Haldane, J.B.S., 'The scientific work of J. S. Haldane', in *Penguin Science Survey*, 1961, 11–33

Haldane, J.B.S., 'A scientist looks into his own grave', in *The Observer Weekend Review*, 10 January 1965

Haldane, J.B.S., 'An autobiography in brief', in *Perspectives in Biology and Medicine*, 9, 1966, 476–481

Haldane, J.S., 'Notes on an enquiry into the nature and physiological action of Black-damp met with in Podmore Colliery, Shropshire', in *Proceedings of the Royal Society of London*, 57, 1895, 249–257

Haldane, J.S., 'Report of a committee appointed by the Lords Commissioners of the Admiralty to consider and report upon the conditions of deep-water diving', in *Parliamentary Paper*, 1549, 1907

Haldane, J.S., 'Memorandum on asphyxiating gases and vapours used by the German troops and on means of protection against them', in NAWO, 142/153 CL/315, 3 May 1915

Haldane, J.S. & J.G. Priestley, *Respiration*. New edition. Clarendon Press, Oxford, 1935

Mitchison, N., *All Change Here: Girlhood and Marriage*, The Bodley Head, London, 1975

Mitchison, N., *You May Well Ask: A Memoir 1920–1940*, Gollancz, London, 1979

Norton, T., 'A history of British diving science', in *Underwater Technology,* 20 (2), 1994, 3–15

Norton, T., 'The absent-minded professor: John Scott Haldane' and 'The cuddly cactus in the chamber of horrors. John Burdon Sanderson Haldane', in *Stars Beneath the Sea*, Arrow Books, London, 2000, 100–118 & 120–143

Norton, T., 'Watch out guinea pigs, here I come', in *Biologist,* 48 (2), 2001, 87–90

Passmore, R. 'The debt of physiologists and miners to J. S. Haldane', in *The Advancement of Science,* 8 (32), 1952, p.418

Pirie, N.W., 'John Burdon Sanderson Haldane' in *Biographical Memoirs of Fellows of the Royal Society,* 12, 1966, 219–249

Sheridan, D., ed., *Among You Taking Notes . . . The Wartime Diaries of Naomi Mitchison,* Gollancz, London, 1985

Warren, C. E. T. & Benson, J., *The Admiralty Regrets . . .,* The Popular Book Club, London, 1958

White, M.J.D., 'J. B. S. Haldane' in *Genetics,* 52, 1965, 1–7

Adrift and Alone

Anon., *Review of the work of the Subcommittee on Protective Clothing of the Associate Committee on Aviation Medical Research 1942–1945,* National Research Council of Canada, Ottawa, June 1946, I–VII + 155 pp.

Ashcroft, F., 'Life in the cold', in *Life at the Extremes,* HarperCollins, London, 2000, 147–183

Bombard, A., *The Bombard Story*, Readers' Union, André Deutsch, London, 1955

Department of Trade, 'Drinking of sea water by castaways', in *Merchant Shipping Notice M–729*, August 1975

Heyerdahl, T., *The Kon-Tiki Expedition*, George Allen & Unwin Ltd, London, 1950

Keating, W.R., *Survival in Cold Water*, Blackwells, London, 1969

Kitching, J.A. & E. Pagé, 'Report to Associate Committee on Aviation Medical Research', in *Subcommittee on Protective Clothing Report* No. 197, 28 July 1945, 1–7

Norton, T., 'Running on treacle. John Alwyne Kitching', in *Stars Beneath the Sea*, Arrow Books, London, 2000, 80–95

Norton, T., *Reflections on a Summer Sea*, Arrow Books, London, 2002

Norton, T., 'Jack of all trades', in *Biologist*, 50 (5), 2003, 236–238

Pain, S., 'Inactive service', in *New Scientist*, 14 December 2002, 52–53

Robin, B., *Survival at Sea*, Stanley Paul & Co Ltd, 1981

Smith, A., *The Body*, Penguin Books, Harmondsworth, revised edition 1985

Stark, B., *Last Breath. Cautionary Tales from the Limits of Human Endurance*, Pan Books, London, 2003

Carnivorous and coming this way

Allen, T.B., *The Shark Almanac*, The Lyons Press, New York, 1999

British Broadcasting Corporation, *Natural world, Great white shark*, BBC2 TV January 2009

Cappuzzo, M., *Close to Shore*, Review, 2002

Clark, E., 'Sharks: Magnificent and misunderstood', in *National Geographic*, February 1981, 138–186

Eibl-Eibesfeldt, I., *Land of a Thousand Atolls*, MacGibbon & Kee, London, 1965

Ferreira, C.A. & T.P. Ferreira, 'Population dynamics of the white shark in South Africa', in A.P. Klimley & D.G. Ainley, eds, *Great White Shark: the Biology of Carcharodon carcharias*, Academic Press, London, 1998, 381–391

Gilbert, P.W., 'The behavior of sharks', in *Scientific American*, July 1962, 2–10

Gilbert, P.W. & S. Springer, 'Testing shark repellents', in Gilbert, P.W., ed., *Sharks and Survival*, Heath & Co., Boston, 1963, 477–494

Gilbert, P.W. & C. Gilbert, 'Sharks and shark repellents', in *Underwater Journal*, 5, April 1973, 69–80

Hass, H. *Diving to Adventure*, Jarrolds, London, 1952

Hass, H. *Under the Red Sea*, Jarrolds, London, 1952

Kenny, N.T., 'Sharks: The wolves of the sea', in *National Geographic*, February 1968, 222–259

Lech, R.B., *The Tragic Fate of the U.S.S. Indianapolis*, Cooper Square Press, New York, 2001

Maniguet, X., *The Jaws of Death*, HarperCollins, London, 1992

Nelson, D.R., R.R. Johnson, J.N. Mckibben & C.G. Pittenger,

'Agonistic attacks on divers and submersibles by gray reef sharks, *Carcharhinus amblyrhynchos*: antipredatory or competitive?' in *Bulletin of Marine Science*, 38, 1986, 68–88

Nelson, D.R. & W.R. Strong, 'Chemical repellent tests on white sharks with comments on repellent delivery', in A.P. Klimley & D.G. Ainley, eds, *Great White Shark: the Biology of Carcharodon carcharias*, Academic Press, London, 1998

Norton, T., 'Diving to adventure. Hans Heinrich Romulus Hass', in *Stars Beneath the Sea*, Arrow Books, London, 2000, 198–216

Taylor, P.L., 'It's all fun and games until someone gets munched', in *Science at the Extreme*, McGraw-Hill, New York, 2001, 204–229

Tricas, T.C. & J.E. McCosker, 'Predatory behavior of the white shark (*Carcharodon carcharias*) with notes on its biology', in *Proceedings of the California Academy of Sciences*, 43, 1984, 221–238

Tuve, R. L., 'Development of the US Navy "Shark Chaser" chemical repellent', in Gilbert, P.W., ed., *Sharks and Survival*, Heath & Co., Boston, 1963, 455–463

Webster, D.K., *Myth and Maneater*, Dell Publishing Co. Ltd., 1975

Into the abyss

Anon, 'Rekord und tod', in *Stern*, December, 1963, 5 pp.

Barak, A., 'The great Scandinavian adventure', in *Historical Diving Times*, 35, 2005, 58–63

Barton, O., *Adventure on Land and Under the Sea*, Longmans, Green & Co., London, 1954

Beebe, W., *Half Mile Down*, The Bodley Head, London, 1935

Bühlmann, A.A., P. Frei & H. Keller, 'Saturation and desaturation with N2 and He at 4 atm.', in *Journal of Applied Physiology*, 23 (4), 1967, 458–462

Craig, J.D., B.K. Hastings, M.C. Degn, H. Bischel & L. Thompson, 'US findings on the fatal dive', in *Triton*, March–April 1963, 25–26

Dugan, J., *Man Explores the Sea*, Pelican Books, Harmondsworth, 1960

Eaton, B., 'Peter Small', in *Triton*, 1974, 258–259

Eaton, B., 'Neptune, Triton, Diver', in *Diver*, July 1993, 36–37

Editorial, 'The man who lived and died – for diving', in *Topic*, 15 December 1962, p. 27

Franzen, A., 'Ghost from the depths: the warship *Vasa*', in *National Geographic*, January 1962, 42–57

Gustafsson, L., 'Zetterström's hydrox experiment', lecture at *Annual Conference of the Historical Diving Society*, Bristol, October 2005

Gustafsson, L., 'Zetterström's hydrox experiment', summary by R. Vallintine in *Historical Diving Times*, 38, 2006, 42–43

Hass, H., *Conquest of the Underwater World*, David & Charles, Newton Abbot, 1975

Honour, A., *Ten Miles High Two Miles Deep*, Brockhampton Press, Leicester, 1959

Keller, H., 'The mistakes at Catalina', in *Triton*, March–April 1963, p. 28

Keller, H., 'Use of multiple inert gas mixtures in deep diving', in C.J. Lambertson, ed., *Underwater Physiology*, Williams & Wilkins, Baltimore, 1967, 267–274

Keller, H. & A.A. Bühlmann, 'Deep diving and short decompression by breathing mixed gases', in *Journal of Applied Physiology*, 20(6), 1965, 1267–1270

Leach, D.L., 'Down to the *Thresher* by bathyscaphe', in *National Geographic*, June 1964, 764–777

Norton, T., 'The delights of dangling. Charles William Beebe', in *Stars Beneath the Sea*, Arrow Books, London, 2000, 54–98

Norton, T., *Reflections on a Summer Sea*, Arrow Books, London, 2002

Piccard, A., *In Balloon and Bathyscaphe*, Cassell & Co. Ltd, London, 1956

Piccard, J. & R.S. Dietz, *Seven Miles Down*, Longmans, Green & Co., London, 1962

Swann, C., 'The development of commercial helium diving', lecture at *Annual Conference of the Historical Diving Society*, Liverpool, October 2008

Throckmorton, P., *The Lost Ships*, Jonathan Cape, London, 1965

Vann, R.D., 'Decompression theory and applications', in P.B. Bennett & D.H. Elliott, *The Physiology and Medicine of Diving*, 3rd edition, Best Publishing Co., San Pedro, 1982, 352–282

Wendling, J., P. Nussberger & B. Schenk, 'Milestones of the

Deep-Diving Research Laboratory, Zurich', in *South Pacific Underwater Medical Sciences*, 29(2), 1999, 91–98

Zetterström, A., 'Djupdykning med syntetiska gasblandningar', in *Teknisk Tidskrift*, 7, 1945, 173–177

High, fast and hazardous

Altman, L. K., *Who Goes First?*, University of California Press, Berkeley, 1998

Ashcroft, F., 'Life at the top', in *Life at the Extremes*, HarperCollins, London, 2000, 5–40

British Broadcasting Corporation, *Rain*, BBC2 TV, 29 April 2009

Collins, P., 'Over Niagara Falls in a barrel of spikes', in *New Scientist*, 12 February 2009, 44–45

DiGiovanni, C. & R.M. Chambers, 'Physiologic and psychologic aspects of the gravity spectrum', in *New England Journal of Medicine*, 270, 1964, 34–41, 88–94, 134–138

Faith, N, *Crash. The Limits of Car Safety*, Boxtree, London, 1997

Franklin, J. & J. Sutherland, *If I Die in the Service of Science*, Morrow, New York, 1984

Green, N.D.C., 'Effects of long-duration acceleration', in Rainford, D.J. & D.P. Gradwell, eds, *Ernstings's Aviation Medicine*, 4th edition, Hodder Arnold, London, 2006, 137–158

Haldane, J.S. & J.G. Priestley, *Respiration*. New edition. Clarendon Press, Oxford, 1935

Hepper, A.E., 'Restraint systems and escape from aircraft', in Rainford, D.J. & D.P. Gradwell, eds, *Ernstings's Aviation Medicine*, 4th edition, Hodder Arnold, London, 2006, 373–384

Honour, A., *Ten Miles High Two Miles Deep*, Brockhampton Press, Leicester, 1959

Howard, P., 'The dangerous deserts of space', in A. Berry, *Harrap's Book of Science Anecdotes*, Harrap, London, 1989, 38–42

Jarret, P., *Pioneer Aircraft*, Putnam, London, 2002

Lewis, M.E., 'Short-duration acceleration', in Rainford, D.J. & D.P. Gradwell, eds, *Ernstings's Aviation Medicine*, 4th edition, Hodder Arnold, London, 2006, 169–177

Lovelace, W.R., 'Physiologic effects of reduced barometric pressure on man', in *Collected Papers of the Mayo Clinic*, 1941, 1–34

Middleton, D., *Test pilots. The Story of British Test Flying 1903–1984*, Guild Publishing, London, 1985

O'Sullivan, D. & D. Zhou, 'Aircrew and cosmic radiation', in Rainford, D.J. & D.P. Gradwell, eds, *Ernstings's Aviation Medicine*, 4th edition, Hodder Arnold, London, 2006, 417–431

Pain, S., 'Higher and higher', in *New Scientist*, 3 July 1999, 52–53

Pain, S., 'The accidental astronaut', in *New Scientist*, 12 September 2007, 54–55

Piccard, A., *In Balloon and Bathyscaphe*, Cassell & Co. Ltd, London, 1956

Stapp, J.P., 'Human tolerance to deceleration', in *American Journal of Surgery*, 93, 1957, 734–740

Stapp, J.P. & W.C. Blout, 'Effects of mechanical force on living tissues II: supersonic deceleration and wind blast', in *Journal of Aviation Medicine*, 27, 1956, 407–416

Whittingham, H.E., 'Medical problems in aviation', in *Chambers Encyclopaedia*, Vol. 2, 1955, 5–8

Yeager, C. & L. Janos, *Yeager. An Autobiography*, Bantam books, New York, 1985

Risky business

Altman, L. K., *Who Goes First?*, University of California Press, Berkeley, 1998

Baggini, J., 'Born to be wild', in *Secret Pioneers, The Observer*, 2008, 39–41

Barnet, A., 'Patients used as drug guinea pigs', in *The Observer*, 9 February 2003, 10–11

Barnet, A., 'UK drug firms used orphans for HIV trials', in *The Observer*, 4 April 2004

Baxby, D., 'The end of smallpox', in *History Today*, March 1999, 14–16

Coleman, V., *Why Animal Experiments Must Stop*, European Medical Journal, Barnstable, 1994

Collins, P., 'Sweet elixir of death', in *New Scientist*, 28 August, 2004, 48–49

Forssmann, W., *Experiments on Myself*, St Martin's Press, New York & London, 1974

George, A., 'Hard to swallow', an interview with Barry

Marshall, in *New Scientist*, 9 December 2006, p. 53

Haldane, J.B.S., 'On being one's own rabbit', in *Possible Worlds*, Chatto & Windus, London, 1927, 107–119

Haldane, J.B.S., *Keeping Cool*, Chatto & Windus, London, 1940

Halpern, S.A., *Lesser Harms. The Morality of Risk in Medical Research*, University of Chicago Press, 2004

Leake, D.C., 'Technical triumphs and moral muddles', in *Annals of Internal Medicine*, 67 (suppl. 7), 1967, 43–56

Lederer, S., *Subjected to Science. Human Experimentation in America before the Second World War*, Johns Hopkins University Press, Baltimore, 1995

Norton, T., 'Living Proof', in *Times Educational Supplement. Curriculum Special*, 2001, 6–7

Osler, W., 'The evolution of the idea of experiment in medicine', in *Transactions of the Congress of Physicians and Surgeons*, 7, 1907, 1–8

Pappworth, M.H., *Human Guinea Pigs. Experimentation on Man*, Pelican Books, Harmondsworth, 1967

Piccard, A., *In Balloon and Bathyscaphe*, Cassell & Co. Ltd, London, 1956

Weyers, W., *The Abuse of Man*, Ardor Scribendi Ltd, New York, 2003

Index

Index

Index